A Handbook of Mathematical Tables and Formulae

A Handbook of Mathematical Tables and Formulae

D. N. GHOSE, MCE, FIE, FIWWA,
FIPHE, MISE, MIAWPC

CBS

CBS Publishers & Distributors Pvt. Ltd.

New Delhi • Bengaluru • Chennai • Kochi • Kolkata • Mumbai
Hyderabad • Nagpur • Patna • Pune • Vijayawada

ISBN: 978-81-239-0910-3

First Edition: 1998
Reprint: 2000, 2002, 2004, 2005, 2007,
2009, 2010, 2012, 2014, 2018

Published by **Satish Kumar Jain** and produced by **Varun Jain** for
CBS Publishers & Distributors Pvt. Ltd.,
4819/XI Prahlad Street, 24 Ansari Road, Daryaganj, New Delhi - 110002
delhi@cbspd.com, cbspubs@airtelmail.in • www.cbspd.com
Ph.: 23289259, 23266861, 23266867 • Fax: 011-23243014

Corporate Office: 204 FIE, Industrial Area, Patparganj, Delhi - 110 092
Ph: 49344934 • Fax: 011-49344935
E-mail: publishing@cbspd.com • publicity@cbspd.com

Branches:
- ***Bengaluru:*** 2975, 17th Cross, K.R. Road, Bansankari 2nd Stage, Bengaluru - 70 • Ph: +91-80-26771678/79 • Fax: +91-80-26771680 E-mail: cbsbng@gmail.com, bangalore@cbspd.com
- ***Chennai:*** No. 7, Subbaraya Street, Shenoy Nagar, Chennai - 600030 Ph: +91-44-26681266, 26680620 • Fax: +91-44-42032115 E-mail: chennai@cbspd.com
- ***Kochi:*** Ashana House, 39/1904, A.M. Thomas Road, Valanjambalam, Ernakulum, Kochi • Ph: +91-484-4059061-65 Fax: +91-484-4059065 • E-mail: cochin@cbspd.com
- ***Kolkata:*** 6-B, Ground Floor, Rameshwar Shaw Road, Kolkata - 700014 Ph: +91-33-22891126/7/8 • E-mail: kolkata@cbspd.com
- ***Mumbai:*** 83-C, Dr. E. Moses Road, Worli, Mumbai - 400018 Ph: +91-9833017933, 022-24902340/41 • E-mail: mumbai@cbspd.com

Representatives:
- Hyderabad: 0-9885175004
- Patna: 0-9334159340
- Jharkhand: 0-9811541605
- Nagpur: 0-9021734563
- Pune: 0-9623451994
- Uttarakhand: 0-9716462459

Printed at:
J.S. Offset Printers, Delhi (India)

Preface

The subject "Mathematics" is one of the fundamental natural sciences which is essentially needed at all times and in every sphere of life. Unquestionably, mathematics is everlasting—as true in the future as it has been in the past.

Sometimes, it becomes difficult to recall the mathematical data and formulae in need, for which such a handbook is essential for ready reference. The present volume contains Mathematical Tables and Formulae. I shall be happy if this volume is granted by all as a handy and reliable source for the mathematical information that will be needed many times in years to come.

—D.N. GHOSE

ABOUT THE AUTHOR

Shri D.N. Ghose, born in 1938, is an well-known author. His first book was published in 1956. Three of his books were selected by the Government of West Bengal (1956–60) for Adult Education. In 1972, National Council of Educational Research & Training, New Delhi, awarded Certificate of Merit for his manuscript "Purano Sei Diner Katha" which ranked second in 17th All India Children's Literature Competition.

Shri Ghose obtained Bachelor Degree (1963) and Master Degree (1969) in Civil Engineering from Jadavpur University. In early 1959, he started his career as a Teacher in Engineering and afterwards held the posts of Head of the Department of Civil Engineering and offg. Principal of Don Bosco Polytechnic, Calcutta.

In 1973, Shri Ghose organised Refresher Course on Public Health Engineering in the Institution of Engineers, India and conducted Unitary Summer School on Public Health Engineering under the sponsorship of India Society for Technical Education. He was also Examiner (1966–72) of various qualifying examinations in the field of Civil Engineering.

Shri Ghose as Public Health Engineer (1973–77) of Tata Engineering & Locomitive Company was responsible for operation, control and maintenance of water treatment plants, sewage treatment plant and town water supply of TELCO. In early 1977, he joined Calcutta Metropolitan Development Authority and visited U.S.A. with a work programme on Computer Analysis of the Water Distribution Network for the City of Calcutta. During his stay in CMDA, he was engaged in planning, design and development work on sewerage, drainage and solid waste management in CMD area. Presently, Shri Ghose is working as a consultant in the field of Environmental Engineering.

Contents

Earth Data

Characteristic	*Value*
Semi-major axis	6·378 388 × 10^6 m
Semi-minor axis	6·356 912 × 10^6 m
Radius of sphere of same volume	6·371221 × 10^6 m
Land area	148·847 × 10^6 Km^2
Ocean area	361·254 × 10^6 Km^2
Mean density of earth	5·522 gm/cm^3
Highest mountain, Everest	8·840 × 10^3 m
Greatest sea depth	10·430 × 10^3 m

Greek Alphabets used in Mathematics

A	α	Alpha	N	ν	Nu
B	β	Beta	Ξ	ξ	Xi
Γ	γ	Gamma	O	o	Omicron
∇	δ	Delta	π	π	Pi
E	ϵ	Epsilon	P	ρ	Rho
Z	ξ	Zeta	Σ	σ	Sigma
H	η	Eta	T	τ	Tau
Θ	θ	Theta	ϒ	ν	Upsilon
I	L	Iota	Φ	ϕ	Phai
K	κ	Kappa	X	x	Chi
Λ	λ	Lambda	ψ	ψ	Psi
M	μ	Mu	Ω	ω	Omega

Mathematical Symbols

Symbol	Meaning	Symbol	Meaning
$+$	Plus/Positive	a^n	*n*th power of a
$-$	Minus/Negative	$\angle$	Any angle
$\pm$	Plus or minus	$\llcorner$	Right angle
$\times$	Multiplied by	$\perp$	Perpendicular to
$\div$	Divided by	$\parallel$	Parallel to
$=$	Equal to	$\triangle$	Triangle
$\neq$	Not equal to	n°	*n* degree
$\cong$	(Equals approx. Congruent)	n'	*n* prime
$>$	Greater than	n''	*n* second
$<$	Less than	$\nparallel$	Not parallel to
$\sim$	Similar to	$\propto$	Varies as
$\geqq$	Greater than or equal to	∞	Infinity
$\leqq$	Less than or equal to	$f(x)$	Function of x
$!$	Factorial	Δx	Increment of x
j or i	Imaginary quantity	$\frac{dy}{dx}$	Derivative of $y = f(x)$ w.r.t. x
$\therefore$	Therefore	dy	Differential of y
$\because$	Since	$\frac{d^2y}{dx^2}$	Second derivative of y
$:$	is to	Sin^{-1}	Anti-Sine
$::$	as to	$\rightarrow$	Approaches as a limit
$\sqrt{}$	Square root	Σ	Summation of
$\sqrt[3]{}$	Cube root	$\int$	Integral of
$\sqrt[n]{}$	*n*th root	$\int_a^b$	Integral between limits a and b

Important Constants

$\pi = 3{\cdot}1416$	$\frac{1}{\pi} = 0{\cdot}3183$
$2\pi = 6{\cdot}2832$	$\frac{1}{2\pi} = 0{\cdot}1591$
$4\pi = 12{\cdot}5664$	$\frac{4}{\pi} = 1{\cdot}2732$
$\frac{\pi}{2} = 1{\cdot}5708$	$\pi^2 = 9{\cdot}8696$
$\frac{\pi}{3} = 1{\cdot}0472$	$\frac{1}{\pi^2} = 0{\cdot}1013$
$\frac{4\pi}{3} = 4{\cdot}1888$	$\sqrt{\pi} = 1{\cdot}7724$
$\frac{\pi}{4} = 0{\cdot}7854$	$\frac{1}{\sqrt{\pi}} = 0{\cdot}5642$
$\sqrt{\frac{4}{\pi}} = 1{\cdot}1284$	$\sqrt[3]{\frac{3}{4\pi}} = 0{\cdot}6203$
$\sqrt[3]{\pi} = 1{\cdot}4646$	$\sqrt[3]{\pi^2} = 2{\cdot}1450$
$\frac{1}{\sqrt[3]{\pi}} = 0{\cdot}6828$	$\log_{10}\pi = 0{\cdot}49715$
$e = 2{\cdot}71828$	$\log_e\pi = 1{\cdot}14473$
$\log_{10}e = 0{\cdot}43429$	$\log_{10}\log_{10}e = \bar{1}.63778$
$\log_e 10 = 2{\cdot}30258$	

Prefix	*Absrevation*	*Scientifiic/Engg. Notation*
Micro mew	$\mu\mu$	10^{-2}
Millimicro	$m\mu$	10^{-9}
Micro	μ	10^{-6}
Milli	m	10^{-3}
Centi	c	10^{-2}
Deci	d	10^{-1}
Deka	da	10
Hecto	h	10^{2}
Kilo	k	10^{3}
Mega	M	10^{6}
Kilomeg.	G	10^{9}
Megamega	T	10^{12}

Algebra

LAWS, FORMULAE, EQUATIONS, EXPENSIONS AND SERIES

Fundamental Laws :

(*i*) Cumulative law : $a+b=b+a$

$ab=ba$

(*ii*) Associative law : $l+(m+n)=(l+m)+n$

$l(mn)=(lm)n$

(*iii*) Dsitributive law : $p(q+r)=pq+pr$

Laws of Exponents :

$a^m \cdot a^n = a^{m+n}$; $(ab)^n = a^n b^n$; $(a^m)^n = a^{mn}$

$x^0 = 1$ if $x \neq 0$; $a^{-n} = \dfrac{1}{a^n}$; $\dfrac{a^m}{a^n} = a^{m-n}$

$a^{m/n} = \sqrt[n]{a^m}$

Operations with Zero :

$m-m=0$; $m \cdot 0 = 0$; $0^a = 0$; $a^0 = 1$; $\dfrac{0}{a} = 0$

if $a \neq 0$

Ratio :

If $\dfrac{a}{b} = \dfrac{c}{d}$, then $\dfrac{b}{a} = \dfrac{d}{c}$ ......Invertendo

$\dfrac{a}{c} = \dfrac{b}{d}$ ......Alternendo

$\dfrac{a+b}{b} = \dfrac{c+d}{d}$ ......Componendo

$\dfrac{a-b}{b} = \dfrac{c-d}{d}$ ......Dividendo

$\dfrac{a+b}{a-b} = \dfrac{c+d}{c-d}$ ......Componendo and dividendo

Laws of Indices :

For rational values of m and n and if a and b are positive

$a^m \times a^n = a^{m+n}$ $(a^m)^n = a^{mn}$ $\sqrt[m]{a} = a^{1/m}$

$\frac{a^m}{a^n}=a^{m-n}$ $(ab)^m=a^m b^m$ $\sqrt[n]{a^m}=a^{m/n}$

$\frac{1}{a^n}=a^{-n}$ $\left(\frac{a}{b}\right)^n=\frac{a^n}{b^n}$ $a^0=1$

Surds :

If $u\pm\sqrt{x}=v\pm\sqrt{y}$, then $u=v$ and $x=y$. u and v are rationals and $\sqrt{x}$ and $\sqrt{y}$ are cailed Surds.

If $\sqrt{u+\sqrt{v}}=\sqrt{x}+\sqrt{y}$,

then $\sqrt{u-\sqrt{v}}=\sqrt{x}-\sqrt{y}$

Complex Numbers :

(a and b are real numbers)

$i=\sqrt{-1}$, $i^2=-1$, $i^3=-1$, $i^4=1$, $i^5=1$, etc.

If $a+ib=c+id$, then $a=c$ and $b=d$

$(a+ib)+(c+id)=(a+c)+i(b+d)$

$(a+ib)\,(c+id)=(ac-bd)+i(ad+bc)$

$$\frac{a+ib}{c+id}=\frac{(a+ib)(c-id)}{(c+id)(c-id)}=\frac{ac+bd}{c^2+d^2}+i\frac{bc-ad}{c^2+d^2}$$

Arithmatic Progression (A.P.) :

$a,\ a+d,\ a+2d,\ a+3d,\ldots\ldots\ldots$to n terms

where a=1st term and d=common difference.

Each term is greater than the previous term by d.

Thus, last term, $l=a+(n-1)d$.

Summation of all the terms upto n is given by

$S_n=a+(a+d)+(a+2d)+\ldots\ldots\ldots+[a+(n-1)d]$.

$=\frac{n}{2}\,(a+l)$.

Arithmatic mean between a and $b=\frac{a+b}{2}$

Geometric Progression (G.P.) :

$a,\ ar,\ ar^2,\ ar^3,\ldots\ldots\ldots\ldots\ldots\ldots$to n

Where a=1st term, r=common ratio, n=number of terms. Then, last term $l=ar^{n-1}$

Sum of all the terms upto n is given by

$$S_n = a\left(\frac{r^n - 1}{r-1}\right) = \frac{rl - a}{r-1} \qquad \text{when } r > 1$$

$$S_n = a\left(\frac{1-r^n}{1-r}\right) \qquad \text{when } r < 1$$

Sum of infinite number of terms of a G.P. is given by

$$S_\infty = \frac{a}{1-r}$$

Geometric mean between a and $b = \sqrt{ab}$

Harmonic Progression :

(Sequence of numbers whose reciprocals form an A.P.)

$$\frac{1}{a}, \frac{1}{a+d}, \frac{1}{a+2d} \text{ ....to } n \text{ terms.}$$

Sum of all the terms upto n is given by

$$S_n = \frac{1}{a} + \frac{1}{a+d} + \frac{1}{a+2d} + \dots\dots\dots \frac{1}{a+(n-1)d}$$

Harmonic mean of a and b is $\frac{2ab}{a+b}$

Means :

If $a_1, a_2, \dots\dots\dots a_n$ are n numbers, then

1. Arithimetic mean $= \frac{1}{n}\left(a_1 + a_2 + \dots\dots + a_n\right)$

2. Geometric mean $= (a_1, a_2 \dots\dots a_n)^{1/n}$

3. Harmonic mean $= n\left(\frac{1}{a_1} + \frac{1}{a_2} + \dots\dots\dots \frac{1}{a_n}\right)^{-1}$

Proportionality or Variation :

Let Constant $= K$. If $x \propto$ directly as y, $x = yK$

If $x \propto$ inversely as y, $x = \frac{K}{y}$

If $x \propto$ jointly as y and z, $x = Kyz$

If $x \propto$ directly as y and inversely as z, $x = K\frac{y}{z}$

Sum of a few Series :

$$\Sigma r = 1+2+3+\ldots\ldots\ldots\ldots+n = \frac{n(n+1)}{2}$$

$$\Sigma r^2 = 1+2^2+3^2+\ldots\ldots\ldots+n^2 = \frac{n(n+1)(2n+1)}{6}$$

$$\Sigma r^3 = 1+2^3+3^3+\ldots\ldots\ldots+n^3 = \left\{\frac{n(n+1)}{2}\right\}^2 = (\Sigma r)^2$$

$$\Sigma(2r-1) = 1+3+5+7+\ldots\ldots = n^2$$

$$\Sigma(2r-1)^2 = 1+3^2+5^2+7^2+\ldots\ldots\ldots+(2n-1)^2 = \frac{n(4n^2-1)}{3}$$

$$\Sigma(2r-1)^3 = 1+3^3+5^3+7^3+\ldots\ldots\ldots+(2n-1)^3 = n^2(2n^2-1)$$

Laws of Logarithms :

If m, n and a are apositive numbers and $a \neq 1$,

$\log_a (mn) = \log_a m + \log_a n$; $\log_a a = 1$; $\log_a 1 = 0$;

$\log_a\left(\frac{m}{n}\right) = \log_a m - \log_a n$; $\log_b a = \frac{1}{\log_a b}$; $\log_a\left(\frac{1}{m}\right) = -\log_a m$

$\log_a m^n = n \log_a m$; $a^b = e^{b \log a}$;

$\log_n m = \frac{\log_a m}{\log_a n}$; $a^{\log_a x} = x$;

$\log_a \sqrt[r]{m} = \frac{1}{r} . \log_a m$; $\log_a m = \log_b m . \log_a b = \frac{\log_b m}{\log_b a}$;
when $b \neq 1$

Permutations : (different arrangement of a a set of things)

$$P(n, r) = {}^nP_r = n(n-1)(n-2)\ldots\ldots\ldots[n-(r+1)] = \frac{n\,!}{(n-r)\,!}$$

$${}^nP_n = n(n-1)(n-2)\ldots\ldots\ldots3.2.1. = n\,!$$

Combinations :

$$C(n,r) = {}_nC_r = \binom{n}{r} = \frac{{}^nP_r}{r\,!} = \frac{n\,(n-1)\,(n-2)\,\ldots\{n-(r+1)\}}{r\,(r-1)}$$

$$= \frac{n\,!}{r\,!\,(n-r)\,!}$$

$$nC_r = nC_{(n-r)}\ ;\quad nC_r + nC_{(r-1)} = (n+1)C_r$$

Expansions and Factors :

$$(a \pm b)^2 = a^2 \pm 2ab + b^2.$$

$$(a \pm b)^3 = a^3 \pm 3a^2b + 3ab^2 \pm b^3.$$

$$(a+b+c)^2 = a^2 + b^2 + c^2 + 2ab + 2bc + 2ca.$$

$$a^2 - b^2 = (a+b)(a-b).$$

$$a^3 + b^3 = (a+b)(a^2 - ab + b^2).$$

$$a^3 - b^3 = (a-b)(a^2 + ab + b^2)$$

$$a^n - b^n = (a-b)(a^{n-1} + a^{n-2}b + \ldots\ldots + b^{n-1})$$

$$a^n - b^n = (a+b)(a^{n-1} - a^{n-2}b + \ldots\ldots\ldots\ldots\ldots - b^{n-1})$$ when n is an even integer.

$$a^n + b^n = (a+b)(a^{n-1} - a^{n-2}b + \ldots\ldots + b^{n-1})$$ when n is an odd integer.

$$a^4 + a^2b^2 + b^4 = (a^2 + ab + b^2)(a^2 - ab + b^2).$$

Binomial theorem and expansion :

$$(a+b)^n = a^n + na^{n-1}b + \frac{n(n-1)}{2!}a^{n-2}b^2 + \frac{n(-1(n-2)}{3!}a^{n-3}b^3 + \ldots\ldots\ldots\ldots + nab^{n-1} + b^n.$$

This expansion may be written in another form ;

$$(a+b)^n = \binom{n}{0}a^n + \binom{n}{1}a^{n-1}b + \binom{n}{2}a^{n-2}b^2 + \binom{n}{3}a^{n-3}b^3 + \ldots\ldots + \binom{n}{n-1}ab^{n-1} + \binom{n}{n}b^v.$$

This can be made short as follows :

$$(a+b)^n = \sum_{r=n}^{u} \binom{n}{r} a^{n-r} b^r$$

Quadratic Equation : If $ax^1 + bx + c = 0$ and $a \neq 0$

then $x = \dfrac{-b \pm \sqrt{b^2 - 4ac}}{2a}$;

If a, b, and c are real and

if $b^2 - 4ac > 0$, the roots are real and unequal :

if $b^2 - 4ac < 0$, the roots are imaginary ;

if $b^2 - 4ac = 0$, the roots are real and equal.

Let $\propto$, and β be the two roots ; Then,

$$\propto = \frac{-b+\sqrt{b^2-4ac}}{2a} \text{ and } \beta = \frac{-b-\sqrt{b^2-4ac}}{2a}$$

Sum of the roots $= -\frac{b}{a}$ and

The product of the roots $= \frac{c}{a}$

If $a=0$, one root is infinite.
If $c=0$, one root is zero.
If $b=c=0$, two root are zero.
If $a=b=0$, two roots are infinite.
If $a=c$, roots are reciprocal.

Cubic Equation :

$$y^3+ly^2+my+n=0$$

Substituting $y=(x-l/3)$, the eqn may be reducd to the normal form $x^3+ax+b=0$;

where $a=\frac{1}{3}(3m-l^2)$ and $b=\frac{1}{27}(2l^3-9lm+27n)$

The solution of the eqn. is as follows :

$x^1=A+B$ and $x'', x''' = -\frac{(A+B)}{2} \pm \frac{i\sqrt{3}}{2}(A-B)$

where $i=\sqrt{-1}$, $A=\sqrt[3]{-\frac{b}{2}+\sqrt{\frac{b^2}{4}+\frac{a^3}{27}}}$, and

$$B=\sqrt[3]{-\frac{b}{2}-\sqrt{\frac{b^2}{4}+\frac{a^3}{27}}}$$

If l, m, and n are real *i.e.*, a and b are real, and

if $\frac{b^2}{4}+\frac{a^3}{27}>0$, one root is real and two are conjugate imaginary roots.

if $\frac{b^2}{4}+\frac{a^3}{27}<0$, three roots are real, and unequal, which is impractical.

if $\frac{b^2}{4}+\frac{a^3}{27}=0$, There roots are real of which at least two are equal.

Biquadratic or Quartic Equation ;

$y^4+py^3+qy^2+ry+s=0$ may be reduced

to $x^4+ax^2+bx+c=0$ by substituing $y=\left(x-\frac{p}{4}\right)$.

Roots are : $x_1=\pm(-\sqrt{l}-\sqrt{m}-\sqrt{n})$

$x_2=\pm(-\sqrt{l}+\sqrt{m}+\sqrt{n})$

$x_3=\pm(+\sqrt{l}-\sqrt{m}+\sqrt{n})$

$x_4=\pm(+\sqrt{l}+\sqrt{m}-\sqrt{n})$

where l, m, and n are the roots of the resolvent cubic eqn.

Exponential, Logarithmic and Trigonometric Series

$$e=1+\frac{1}{1!}+\frac{1}{2!}+\frac{1}{3!}+\frac{1}{4!}+\ldots\ldots\ldots\ldots$$

$$\frac{1}{e}=1-\frac{1}{1!}+\frac{1}{2!}-\frac{1}{3!}+\frac{1}{4!}-\ldots\ldots\ldots\pm\frac{(-1)^n}{n!}\mp\ldots\ldots\ldots\ldots$$

$$e^x=1+x+\frac{x^2}{2!}+\frac{x^3}{3!}+\frac{x^4}{4!}+\ldots\ldots\ldots\ldots$$

$$a^x=1+x\log a+\frac{(x\log a)^2}{2!}+\frac{(x\log a)^2}{3!}+\ldots\ldots\ldots\ldots$$

$$e^{-x^2}=1-x^2+\frac{x^4}{2!}-\frac{x^6}{3!}+\frac{x^8}{4!}-\ldots\ldots\ldots\ldots$$

$$\log x=(x-1)-\tfrac{1}{2}(x-1)^2+\tfrac{1}{3}(x-1)^3-\ldots\ldots\ldots,\ \text{when } 0<x\leqq 2$$

$$\log x=\frac{x-1}{x}+\frac{1}{2}\left(\frac{x-1}{x}\right)^2+\frac{1}{3}\left(\frac{x-1}{x}\right)^3+\ldots\ldots\ldots\text{when } x>\frac{1}{2}$$

$$\log x=2\left\{\frac{x-1}{x+1}+\frac{1}{3}\left(\frac{x-1}{x+1}\right)^3+\frac{1}{5}\left(\frac{x-1}{x+1}\right)^5+\ldots\ldots\ldots\ldots\right\}\text{when } x>0$$

$$\log(a+x)=\log a+2\left\{\frac{x}{2a+x}+\frac{1}{3}\left(\frac{x}{2a+x}\right)^3+\frac{1}{5}\left(\frac{x}{2a+x}\right)^5+\ldots\right\}$$

when, $a>0$, $-a<x<\propto$

$$\log(1+x)=x-\frac{x^2}{2}+\frac{x^3}{3}-\frac{x^4}{4}+\ldots\ldots\ldots\text{where} -1<x\leqq 1$$

$$\log\left(\frac{1+x}{1-x}\right)=2\left(x+\frac{x^3}{3}+\frac{x^5}{5}+\ldots\ldots\ldots\ldots\right)\text{ when } x^2<1$$

$$\log\left(\frac{x+1}{x-1}\right)=2\left\{\frac{1}{x}+\frac{1}{3}\left(\frac{1}{x}\right)^3+\frac{1}{5}\left(\frac{1}{x}\right)^5+\ldots..\right\} \text{ where } x^2>1$$

$$\log(x+\sqrt{1+x^2})=x-\tfrac{1}{2}\frac{x^3}{3}+\frac{1.3}{2.4}\ \frac{x^5}{5}-\frac{1.3.5}{2.4.6}\ \frac{x^7}{7}+\ldots\ldots \text{when } x^2<1$$

$$\log e^2=1-\frac{1}{2}+\frac{1}{3}-\frac{1}{4}+\ldots\ldots\ldots+(-1)^n\ldots\ldots\ldots$$

$$\text{Sin } x=x-\frac{x^3}{3!}+\frac{x^5}{5!}-\frac{x^7}{7!}+\ldots\ldots\ldots\ldots\ldots\ldots$$

$$\text{Cos } x=1-\frac{x^2}{2!}+\frac{x^4}{4!}-\frac{x^6}{6!}+\ldots\ldots\ldots\ldots\ldots$$

$$\tan x=x+\frac{x^3}{3}+\frac{2x^5}{15}+\frac{17x^7}{315}+\ldots\ldots \text{when } x^2<\frac{\pi^2}{4}$$

$$\log \text{ Sin } x=\log x-\frac{x^2}{6}-\frac{x^4}{180}-\frac{x^6}{2835}-\ldots\ldots\ldots \text{when } x^2<\pi^2$$

$$\log \text{ Cos } x=-\frac{x^2}{2}-\frac{x^4}{12}-\frac{x^6}{45}-\ldots\ldots\ldots \text{when} x^2<\frac{\pi^2}{4}$$

$$\log \tan x=\log x+\frac{x^2}{3}+\frac{7x^4}{90}+\ldots\ldots \text{when } x^2<\frac{\pi^2}{4}$$

$$\text{Sin}^{-1}x=x+\frac{x^3}{6}+\frac{1}{2}\cdot\frac{3}{4}\cdot\frac{x^5}{5}\cdot+\frac{1}{2}\cdot\frac{3}{4}\cdot\frac{5}{6}\cdot\frac{x^7}{7}+\ldots\ldots\ldots\ldots\ldots$$
when $x^2<1$

$$\tan^{-1}x=\frac{\pi}{2}-\frac{1}{x}+\frac{1}{3x^3}-\frac{1}{5x^5}+\ldots\ldots\ldots \text{when } x^2>1$$

$$\tan^{-1}x=x-\frac{1}{3}x^3+\frac{1}{5}x^5-\frac{1}{7}x^7+\ldots\ldots\ldots \text{when } x^2<1$$

$$\text{Sin}h\ x=x+\frac{x^3}{3!}+\frac{x^5}{5!}+\frac{x^7}{7!}+\ldots\ldots\ldots\ldots\ldots$$

$$\text{Cos}h\ x=1+\frac{x^2}{2!}+\frac{x^4}{4!}+\frac{x^6}{6!}+\ldots\ldots\ldots\ldots\ldots$$

$$\tan h\ x=x-\frac{x^3}{3}+\frac{2x^5}{15}-\frac{17x^7}{315}+\ldots\ldots\ldots\ldots$$

$$\text{Sin}h^{-1}x=\log 2x+\tfrac{1}{2}\frac{1}{2x^2}-\frac{1.3}{2.4}\ \frac{1}{4x^4}+\frac{1.3.5}{2.4.6}\ \frac{1}{6x^6}+\ldots.. \text{when } x>1$$

$$\text{Sin}h^{-1}\ x=x-\tfrac{1}{2}\frac{x^3}{3}+\frac{1.3}{2.4}\ \frac{x^5}{5}-\frac{1.3.5}{2.4.6}\ \frac{x^7}{7}+\ldots\ldots\ldots \text{when } x^2<1$$

$$\text{Cos}h^{-1}x=\log 2x-\tfrac{1}{2}\frac{1}{2x^2}-\frac{1.3}{2.4}\frac{1}{4x^4}-\frac{1.3.5}{2.4.6}\frac{1}{6x^6}-\ldots\ldots\ldots\ldots$$

$$\tan h^{-1} x=x+\frac{x^3}{3}+\frac{x^5}{5}+\frac{x^7}{7}+\ldots\ldots\ldots\ldots$$

$$e^{\sin x}=1+x+\frac{x^2}{2!}-\frac{3x^4}{4!}-\frac{8x^5}{5!}-\frac{3x^6}{6!}+\ldots\ldots\ldots\ldots$$

$$e^{\cos x}=e\left\{1-\frac{x^2}{2!}+\frac{4x^4}{4!}-\frac{31x^6}{6!}+\ldots\ldots\right\}$$

$$e^{\tan x}=1+x+\frac{x^2}{2!}+\frac{3x^3}{3!}+\frac{9x^4}{4!}+\ldots\ldots\text{when } x^2<\frac{\pi^2}{4}$$

A few important Series :

$$1+\frac{1}{2}+\frac{1}{4}+\frac{1}{8}+\ldots\ldots\ldots+\frac{1}{2n}+\ldots\ldots=2$$

$$1-\frac{1}{2}+\frac{1}{4}-\frac{1}{8}+\ldots\ldots\ldots\pm\frac{1}{2n}\mp\ldots\ldots=2/3$$

$$1-\frac{1}{3}+\frac{1}{5}-\frac{1}{7}+\frac{1}{9}-\ldots\ldots\pm\frac{1}{2n-1}\pm\ldots\ldots=\pi/4$$

$$\frac{1}{1.2}+\frac{1}{2.3}+\frac{1}{3.4}+\ldots\ldots+\frac{1}{n(n+1)}+\ldots\ldots=1$$

$$\frac{1}{1.3}+\frac{1}{2.4}+\frac{1}{3.5}+\ldots\ldots+\frac{1}{n(n+2)}+\ldots\ldots=3/4$$

$$\frac{1}{1.3}+\frac{1}{3.5}+\frac{1}{5.7}+\ldots\ldots+\frac{1}{(2n-1)(2n+1)}+\ldots\ldots=1/2$$

$$\frac{1}{1.2.3}+\frac{1}{2.3.4}+\ldots\ldots\ldots+\frac{1}{n(n+1)(n+2)}+\ldots\ldots=1/4$$

Vector Analysis :

$V=ai+bj+cK$ where $V=a$ vector in space,

$|V|=+\sqrt{a^2+b^2+c^2}$ $a, b, c=$magnitudes of the projections of V along three axes.

$i, j, k=$vectors of unit magnitude

$|V|=$magnitude of vactor V.

Vector sum of n Vectors ;

$V_1, V_2, V_3,\ldots\ldots\ldots$to n vector

$V_1=a_1i+b_1j+c_1k$, etc.

$$V = V_1 + V_2 + V_3 + \ldots\ldots\ldots\ldots + V_n$$
$$= (a_1 + a_2 + a_3 + \ldots\ldots + a_n)i + (b_1 + b_2 + b_3 + \ldots\ldots\ldots b_n)j$$
$$+ (c_1 + c_2 + c_3 + \ldots\ldots\ldots + c_n)k$$

Scalar Product of two Vectors :

$V_1 \,.\, V_2 = |V_1| \; |V_2| \text{ Cos } \phi$ where ϕ = angle between V_1 and V_2.

$$V_1 \,.\, V_2 = a_1a_2 + b_1b_2 + c_1c_2 = V_2 \,.\, V_1$$

$$V_1 \,.\, V_1 = |V_1|^2$$

$$(V_1 + V_2) \,.\, V_3 = V_1 \,.\, V_3 + V_2 \,.\, V_3$$

$$i \,.\, i = j \,.\, j = k.\, k. = 1$$

$$i \,.\, j = j \,.\, k = k \,.\, i = 0$$

Vector product of two Vectors :

$V_1 \times V_2 = |V_1| \, V_2| \text{ Sin}\phi$, where ϕ = angle between V_1 and V_2

$$V_1 \times V_2 = -V_2 \times V_1 = (b_1c_2 - b_2c_1)i + (c_1a_2 - c_2a_1)j + (a_1b_2 - a_2b_1)k$$

$$(V_1 + V_2) \times V_3 = V_1 \times V_3 + V_2 \times V_3$$

$$V_1 \times (V_2 \times V_3) = V_2(V_1 . V_3) - V_3(V_1 V_2)$$

$$i \times i = j \times j = k \times k = 0$$

$$i \times j = k, \quad j \times k = i, \quad k \times i = j$$

$$V_1 \,.\, (V_2 \times V_3) = (V_1 \times V_2) \,.\, V_3 = V_2 \,.\, (V_3 \times V_1) = [V_1 \; V_2 \; V_3]$$

$$[V_1 \; V_2 \; V_2] = \begin{vmatrix} a_1 & a_2 & a_3 \\ b_1 & b_2 & b_3 \\ c_1 & c_2 & c_3 \end{vmatrix}$$

Differentiation of Vectors :

$$\frac{d}{dt}(V_1 + V_2) = \frac{dV_1}{dt} + \frac{dV_2}{dt}$$

where $\dfrac{dV_1}{dt} = \dfrac{da_1}{dt}\, i + \dfrac{db_1}{dt}\, j + \dfrac{dc_1}{dt}\, k$, etc.

$$\frac{d}{dt}(V_1 \,.\, V_2) = \frac{dV_1}{dt} \,.\, V_2 + V_1 . \frac{dV_2}{dt}.$$

$$\frac{d}{dt}(V_1 \times V_2) = \frac{dV_1}{dt} \times V_2 + V_1 \times \frac{dV_2}{dt}.$$

$$V.\frac{dV}{dt}=|V|\frac{d|V|}{dt}.$$

$$V.\frac{dV}{dt}=0 \text{ when } |V| \text{ is constant.}$$

$$\text{grad } s\equiv\nabla s=\frac{\delta s}{\delta x}i+\frac{\delta s}{\delta y}j+\frac{\delta s}{\delta z}k$$

$$\text{Divergence } V=\nabla.V=\frac{\delta a}{\delta x}+\frac{\delta b}{\delta y}+\frac{\delta c}{\delta z}$$

$$\text{curl } V\equiv\text{rot } V\equiv\nabla\times V$$

$$\nabla^2V=i\nabla^2a+j\nabla^2b+k\nabla^2c$$

$$\text{curl grad } s=0\;;\quad \text{Div. curl } V=0$$

$$\text{curl curl } V=\text{grad div. } V-\nabla^2v$$

Green's Formula :

$$\iiint div\ F\ dv=\iiint\nabla.\ F\ dv=\iint F.\ ds$$

Stoke's Formula :

$$\int F.\ dr=\iint \text{curl } F.\ ds=\iint \nabla\times F.\ ds$$

DETERMINANTS

Determinant of 2nd order :

$$\left.\begin{array}{l}\alpha_1x+\beta_1y=0\\ \alpha_2x+\beta_2y=0\end{array}\right\}\text{ may be written as }\begin{vmatrix}\alpha_1 & \beta_1\\ \alpha_2 & \beta_2\end{vmatrix}=0$$

Determinant of 3rd order :

$$\left.\begin{array}{l}\alpha_1x+\beta_1y+\gamma_1z=0\\ \alpha_2x+\beta_2y+\gamma_2z=0\\ \alpha_3x+\beta_3y+\gamma_3z=0\end{array}\right|\text{ may be written as }\begin{vmatrix}\alpha_1 & \beta_1 & \gamma_1\\ \alpha_2 & \beta_2 & \gamma_2\\ \alpha_3 & \beta_3 & \gamma_3\end{vmatrix}=0$$

where $\alpha_1\ \beta_1\ \gamma_1$ is a row and $\begin{array}{c}\alpha_1\\ \alpha_2\\ \alpha_3\end{array}$ is a column.

Rule of Sarrus (to determine the signs of terms in the expansion of a third order determinant):

$$\text{Let the determinant be } \triangle=\begin{vmatrix}\alpha_1 & \beta_2 & \gamma_1\\ \alpha_2 & \beta_2 & \gamma_2\\ \alpha_3 & \beta_3 & \gamma_3\end{vmatrix}$$

Then, write down as follows :

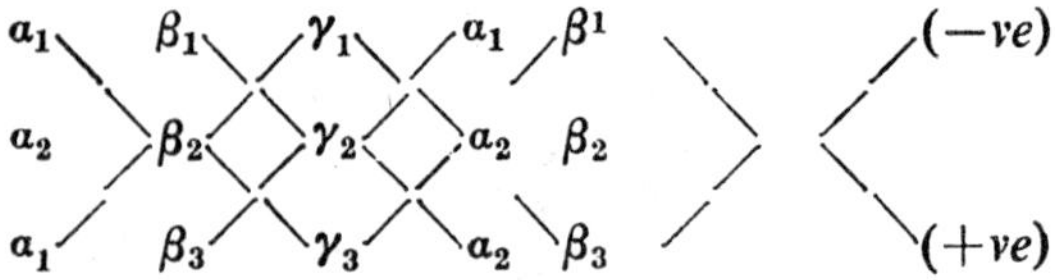

$$\therefore \quad \triangle = \alpha_1 \beta_2 \gamma_3 + \beta_1 \gamma_2 \alpha_3 + \gamma_1 \alpha_2 \beta_3 - \alpha_3 \beta_2 \gamma_1 - \beta_3 \alpha_1 \gamma_2 - \alpha_1 \beta_2 \gamma_3$$

Expansion of a determinant :

Refer. to the above determinant,

$$\triangle = \alpha_1 (\beta_2\gamma_3 - \beta_3\gamma_2) + \beta_1(\gamma_2\alpha_3 - \gamma_3\alpha_2) + \gamma_1(\alpha_2\beta_3 - \alpha_3\beta_2)$$

or,
$$\triangle = \alpha_1 \begin{vmatrix} \beta_2 & \gamma_2 \\ \beta_3 & \gamma_3 \end{vmatrix} + \beta_1 \begin{vmatrix} \alpha_1 & \gamma_2 \\ \alpha_3 & \gamma_3 \end{vmatrix} + \gamma_1 \begin{vmatrix} \alpha_2 & \beta_2 \\ \alpha_3 & \beta_3 \end{vmatrix}$$

Properties of determinant :

1. If we change the columns into rows and the rows into columns, the value of the determinant will remain unaltered.

i.e.,
$$\begin{vmatrix} \alpha_1 & \beta_1 & \gamma_1 \\ \alpha_2 & \beta_2 & \gamma_2 \\ \alpha_3 & \beta_3 & \gamma_3 \end{vmatrix} = \begin{vmatrix} \alpha_1 & \alpha_2 & \alpha_3 \\ \beta_1 & \beta_2 & \beta_3 \\ \gamma_1 & \gamma_2 & \gamma_3 \end{vmatrix}$$

(2) The value of the determinant is not altered, but the sign of the determinant is changed, when two adjacent rows or columns are interchanged.

i.e.,
$$\begin{vmatrix} \alpha_1 & \beta_1 & \gamma_1 \\ \alpha_2 & \beta_2 & \gamma_2 \\ \alpha_3 & \beta_3 & \gamma_3 \end{vmatrix} = - \begin{vmatrix} \alpha_1 & \beta_1 & \gamma_1 \\ \alpha_3 & \beta_3 & \gamma_3 \\ \alpha_2 & \beta_2 & \gamma_2 \end{vmatrix}$$

(3) When any two columns are interchanged, the sign of the determinant is changed, but the numerical value is not altered.

(4) If two rows or columns are identical, the value of the determinant is zero.

i.e.,
$$\begin{vmatrix} \alpha_1 & \beta_1 & \beta_1 \\ \alpha_2 & \beta_2 & \beta_2 \\ \alpha_3 & \beta_3 & \beta_3 \end{vmatrix} = 0$$

(5) If each element in a row or a columnt is multiplied by the same constant, the value of the determinant is multiplied by that constant.

$$\begin{vmatrix} m\alpha_1 & n\beta_1 & \gamma_1 \\ m\alpha_2 & n\beta_2 & \gamma_2 \\ m\alpha_3 & n\beta_3 & \gamma_2 \end{vmatrix} = mn \begin{vmatrix} \alpha_1 & \beta_1 & \gamma_1 \\ \alpha_2 & \beta_2 & \gamma_2 \\ \alpha_3 & \beta_3 & \gamma_3 \end{vmatrix}$$

(6) If each element of a row or column of determinant consists of two terms, the determinant is expressed as the sum of two determinants.

i.e.,
$$\begin{vmatrix} \alpha_1+a_1 & \beta_1 & \gamma_1 \\ \alpha_2+a_2 & \beta_2 & \gamma_2 \\ \alpha_3+a_3 & \beta_3 & \gamma_3 \end{vmatrix} = \begin{vmatrix} \alpha_1 & \beta_1 & \gamma_1 \\ \alpha_2 & \beta_2 & \gamma_2 \\ \alpha_3 & \beta_3 & \gamma_3 \end{vmatrix} + \begin{vmatrix} a_1 & \beta_1 & \gamma_1 \\ a_2 & \beta_2 & \gamma_2 \\ a_3 & \beta_3 & \gamma_3 \end{vmatrix}$$

(7) If each element of *a* row (or column) is diminished or increased by the same multiple of another row (or column) the value of the determinant reamains the same.

(8) If the elements of a determinant are polynomials in a variable x and the determinant vanishes when $x=\lambda$, then $(x-\lambda)$ is a factor of the determinant.

Multiplication of two determinants :

The product of two determinants of any order is itself a determinant of the same order.

Properties of reciprocal determinants :

(1) Let $\triangle$=a determinant of *n*th order, and

$\triangle^1$=its reciprocal determinant

Then, $\triangle^1=\triangle^{n-1}$

(2) Any minor of $\triangle^1$ can be expressed in terms of the elements of $\triangle$.

Symmetric determinants :

A determinant in which every two conjugate elements are equal to each other is called a symmetric determinant.

e.g., $\triangle = \begin{vmatrix} a_{11} & a_{12} & a_{13} \ldots\ldots a_{1n} \\ a_{21} & a_{22} & a_{23} \ldots\ldots a_{2n} \\ a_{n1} & a_{n2} & a_{n3} \ldots\ldots a_{nn} \end{vmatrix}$ Here, a_{12} and a_{21} or, a_{23} and a_{32} are said to be conjugate elements.

Properties of symmetric determinants :

(1) The co-factors of any two conjugate elements in a symmetric determinant are equal.

(2) The square of determinant is a symmetric determinant.

(3) The leading minors of a symmetric determinat are symmetrical determinants :

Skew-symmetric determinant :

When each of the constituents of a determinant equals its conjugate constituent with sign changed, the determinant is said to be skew-symmetric. For example, a determinat

$\triangle \equiv (a_{11}\ a_{22}\ a_{33} \ldots\ldots a_{nn})$ is a skew determinant when

$a_{rs} = -a_{sr},\ r \neq s,\ rs = 1, 2, 3, \ldots\ldots n.$

Propertics of skew-symmetric determinants :

(1) $A_{rs} = (-1)^{n-1} A_{sr}$ where A_{rs} and A_{sr} are cofactors of the conjugate elements a_{rs} and a_{sr} of the determinant.

(2) The value of a skew-symmetric determinant of odd order is zero.

(3) A skew-symmetric determinant of even order is a perfect square.

(4) The square of any determinant of even order can be expressed as a skew-symmetric determinant.

(5) The reciprocal of a skew-symmetric determinant of even order is skew-symmetric determinent.

The product of two reciprocal determinants is the reciprocal determinant of the product of the two original determinants.

MATRICES

When a system of mn numbers arranged in an order set of m rows, each row containing an ordered set of n members, it is called m by n matrix. Each of the mn numbers is called an element. For example, in the matrix

$$\begin{vmatrix} a_{11} & a_{12} & a_{13} \ldots\ldots\ldots a_{1n} \\ a_{21} & a_{22} & a_{23} \ldots\ldots\ldots a_{2n} \\ a_{31} & a_{32} & a_{33} \ldots\ldots\ldots a_{3n} \\ \ldots & \ldots & \ldots \ldots\ldots\ldots\ldots \\ a_{m1} & a_{m2} & a_{m3} \ldots\ldots a_{mn} \end{vmatrix}$$

numbers or functions $a_i j$ are called its elements.

i stands for row and j stands for colunm.

Square matrices: When $m=n$, it is called a square matrix of order n. In a square matrix, the elements a_{1j}, a_{22}, a_{33},......a_{nn} are called diagonal elements.

Equal matrices : Two matrices are said to be equat if and only if they have the same order and each element of one equal to the corresponding element of the other.

Zero matrix : A matrix having all zero elements is called a zero matrix.

Sums of matrices :

$A=[a_{1j}]$ and $B=[b_{1j}]$

$A \pm B=[a_{1j} \pm b_{1j}]=[c_{1j}]=C$, say.

Note that, two matrices of different orders can not be added or subtracted. Assuming matrices $A,B,C,$ are conformable for addition.

$A+B=B+A$

$A+(B+C)=(A+B)+C$

$K\,(A+B)=KA+KB=(A+B)\,K$

Product of Matrices :

If $[a_{1j}]=m \times p$ matrix A and $[b_{1j}]=p \times n$ matrix B then, AB is $m \times n$. By the product AB is meant the $m \times n$ matrix $C=[c_{1j}]$

Assuming that the matries A, B, and C are conformable for sums and products,

$A(B+C)=AB+AC$	$AB \neq BA$ usually
$(A+B)C=AC+BC$	$AB=AC$ does not imply $B=C$
$A(BC)=(AB)C$	$AB=0$ does not imply $A=0$ or $B=0$

Diagonal matrix :

$$\begin{vmatrix} a_{11} & 0 & 0 & \ldots\ldots 0 \\ 0 & a_{22} & 0 & \ldots\ldots 0 \\ 0 & 0 & a_{33} & \ldots\ldots 0 \\ 0 & 0 & 0 & \ldots\ldots 0 \\ 0 & 0 & 0 & \ldots\ldots a_{nn} \end{vmatrix} = \text{diagonal } (a_{11}, a_{22}, a_{33}, \ldots\ldots a_{nn})$$

Scalar Matrix :

If $a_{11}=a_{22}=a_{33}=\ldots\ldots a_{nn}$, it is called a scalar matrix.

Idenlity Matrix :

If $a_{11}=a_{22}=a_{33}=\ldots\ldots=a_{nn}=1$, it is the Identity matrix.

Inverse of a Matrix :

Let A and B be square matrices such that $AB=BA=1$.

Then, B is called the inverse of A and vice-versa.

$B=A^{-1}$ and $A=B^{-1}$.

If two square matrics of the same order A and B have inverses A^{-1} and B^{-1} respectively, then $(AB)^{-1}=B^{-1}.A^{-1}$. This means the inverse of the product of the two matrices having inverses is the product in reverse order of these inverses.

Transpose of a Matrix :

The transpose of a matrix is obtained by interchanging the rows and columms. Thus, the transpose of

$$A=\begin{vmatrix} a_1 & b_1 & c_1 \\ a_2 & b_2 & c_2 \end{vmatrix} \text{ is } A'=\begin{vmatrix} a_1 & a_2 \\ b_1 & b_2 \\ c_1 & c_2 \end{vmatrix}$$

The transpose of the sum of two matrices is the sum of their transposes, *i. e.*, $(A+B)'=A'+B'$

If A' and B' are the transposes of A and B respectively and K is a scalar, then $(KA)'=KA'$ and $(A')'=A$

The transpose of the product of two matrices is the product in reverse order of their transposes, *i.e.*. $(AB)'=B.'A'$.

Symmetric Matrices :

A square matrix $B=[b_{ij}]$ is called symmetric when $b_{ij}=b_{ji}$ for all values of i and j.

For example, let $B=[b_{ij}]$ and $B'=[b_{ji}]$; Then, $B=B'$.

If B is an n square mat ix, then $B+B'$ is symmetric.

If B is any n square matrix, then $B-B'$ is skew-symmetric.

Every square matrix B can be written as the sum of a symmetric matrix $C \frac{B+B'}{2}$ and a skew-symmetric matrix $D=\frac{B-B'}{2}$.

Conjugate of a Matrix :

When B is a matrix having complex numbers as elements, the matrix obtained from B by replacing each element by its conjugate is called the conjugate of B and is denoted by $\overline{B}$

The conjugate of the sum of two matrices is the sum of their conjugates. For example, $(\overline{C+D})=\overline{C}+\overline{D}$

The conjugate of the product of two matrices is the product in the same order of their conjugates. For example, $(\overline{CD})=\overline{C}\ \overline{D}$.

Hermitian Matrices :

A square matrix $B=[b_{ij}]$ such that $\overline{B}'=B$ is called Hermitian. Thus, $b_{ij}=b_{ji}$ for all values of i and j. The diagonal elements of a Hermitian matrix are real numbers :

If B is an n-square matrix, then $B+\overline{B}'$ is Hermitian and $B-\overline{B}°$ is skew-Hermitian.

Matrice Polynomial :

It is an expression of the form

$F(x)=A_0+A_1x+A_2x^2+.........+A_{m-1}x^{m-1}+A_mx^m$.

Where $A_0, A_1, A_2.........A_m$ are all square matrices of the same order and this is of degree m, provided $A_m \neq 0$.

Equivalent Matrices :

Two matrices are said to be equivalent, if one can be obtained from the other by a sequence of elementary tranformations. Thus, equivalent matrices have the same order and the same rank.

Row Equivalent

If a matrix B is reduced to C by using elementary row transformations alone, C is said to be row equivalent to B and vice-versa.

Rank of a Matrix :

A zero matrix is said to have rank zero. *A* non-zero matrix is said to have rank r, if at least one of its r-square minors is different from zero while every $(r+1)$—square minor is zero.

The rank of the product of two matrics can never exceed the rank of either factor.

Adjoint of a Square Matrix :

Let $B=[b_{ij}]$ be an n-square matrix and B_{ij} be the co-factor of b_{ij}.

Then, adj. $B=\begin{vmatrix} B_{11} & B_{21} & \cdots & B_{n1} \\ B_{12} & B_{22} & \cdots & B_{n2} \\ \cdots & \cdots & \cdots & \cdots \\ B_{1n} & B_{2n} & & B_{nn} \end{vmatrix}$

adj $AB=$adj B. adj A

2

Mensuration

1. **Rectangles :** Area=length × breadth.

   $$\text{Diagonal}=\sqrt{(\text{length})^2+(\text{breadth})^2}$$

2. **Squares :** Area=(Side)^2.

   Diagonal=Side $\sqrt{2}$.

3. **Triangles :** Area=$\frac{1}{2}$ base × altitude.

   Also, area=$\sqrt{s(s-a)(s-b)(s-c)}$

   where a, b, and c are the three sides,

   and $S=\dfrac{a+b+c}{2}$

4. **Equilateral Triangles :**

   $$\text{Height}=\frac{\text{Side}\sqrt{3}}{2}$$

   $$\text{Area}=\frac{\sqrt{3}}{4}(\text{Side})^2$$

5. **Right-angled Triangles :**

   $$\text{Hypotenuse}=\sqrt{(\text{base})^2+(\text{perpendicular})^2}$$

   $$\text{Base}=\sqrt{(\text{hypotenuse})^2-(\text{perpendicular})^2}$$

   $$\text{Perpendicular}=\sqrt{(\text{hypotenuse})^2-(\text{base})^2}$$

6. **Isosceles Triangles :**

   $$\text{Area}=\frac{\text{base}}{4}\sqrt{4(\text{Side})^2-(\text{base})^2}.$$

7. **Parallelograms :**

   Area=base × height=diagonal × offset of diagonal

   $=2\sqrt{s(s-a)(s-b)(s-d)}$, where $s=\dfrac{a+b+d}{2}$

   d=diagonal and a. b are two adjacent sides.

8. **Rhombus :** Area$=\frac{1}{2}d_1d_2$, where d_1 and d_2 are two diagonals.

9. **Quadrilaterals :** Area$=\frac{1}{2}d\,(p_1+p_2)$, where d is the diagonal and p_1, p_2 are offsets of the diagonal.

10. **Trapezoids :** Area$=\frac{1}{2}h\,(a+b)$; where a, b are parallel sides and h is the perpendicular distance between the parallel sides.

11. **Regular Hexagons :**

$$\text{Area}=\frac{3\sqrt{3}}{2}\,(\text{Side})^2$$

12. **Regular Octagons :** Area$=4{\cdot}84\,(\text{Side})^2$

13. **Regular Polygions :**

$$\text{Area}=\frac{n}{2}ar.$$ where a=side, n=number of sides,

$$=R^2\times\frac{n}{2}\,\text{Sin}\,\frac{360°}{n}$$ r=radius of inscribed circle,

and R=radius of circumscribed circle.

$$=r^2\times n\tan\frac{180°}{n}.$$

$$=\frac{na}{2}\sqrt{R^2-\left(\frac{a}{2}\right)^2}$$

14. **Circles :** Circumference$=\pi\times$diameter of the circle.

$$\text{Area}=\frac{\pi d^2}{4}\text{ or }\pi r^2$$ where d=diameter and

r=radius of the circle.

15. **Ellipses :** Area$=\pi ab$

where a=semi-major axis, and

b=semi-minor axis

16. **Chords of Circles :**

Semi-chord of the arc$=\sqrt{h(d-h)}$

Chord of the semi-arc$=\sqrt{dh}$

where, d=diameter of the circle,

and h=height of the arc.

17. Arcs of Circles :

Length of the arc=Circumference of the circle$\times\frac{\infty}{360}$

$=\frac{8b-2a}{3}$ where $\propto$ =central angle of the arc in degrees,

a—semi-chord of the arc,

and b=chord of the semi-arc

18. Sectors of Circles :

Area=$\frac{1}{2}\times$ length of the arc$\times$ radius of the circle.

$=\frac{\propto}{360}\times\pi r^2$

19. Segments of Circles : Segment=Sector—triangle formed

Area= $\frac{4}{3}h\sqrt{\frac{c^2}{4}+\frac{2}{5}}h^2$ where, h=height of the segment,

and c=chord of the segment.

20 Rectangular solids :

Diagonal, $d=\sqrt{l^2+b^2+h^2}$ where l=length,

Volume=$l.b.h.$ b=breadth,

h=height or depth,

$=\sqrt{A_1A_2A_3}.$ A_1=area of base,

whole surface A_2=area of side, and

$=2(lb+bh+hl)$ A_3=area of end of the solid.

21. Cubes :

Diagonal=$\sqrt{3}\times$edge,

Volume=(edge)^3 ; whole surface=6$\times$(edge)^2

22. Prisms and Cylinders :

Volume=base area$\times$ height.

=area of cross-section$\times$ length.

23. Circular Cylinders :

Volume=area of base$\times$ height.

24. Right-circular Cylinders :

Volume=base area×height

whole surface=circumference at base (radius of base+height)

25. Pyramids and Cones :

Volume=$\frac{1}{3}$×base area×height.

26. Right-regular Pyramids :

Volume=$\frac{1}{3}$×area of base×height.

whole surface=$\frac{1}{2}$×(perimeter of base×slant height)+Area of base

27. Circular Cones :

Volume=$\frac{1}{3}$×base area×height

28. Right-circular Cones :

Volume=$\frac{1}{3}$×base area×height.

whole surface=$\pi r\times(\sqrt{h^2+r^2}+r)$

where r=radius of base and h=height.

29. Cylindrical Rings :

Volume $\frac{\pi^2}{4}(R+r)(R-r)^2$ where, R=outer radius

whole surface=$\pi^2(R^2-r^2)$. r=inner radius

30. Regular Tetrahedrons :

Volume $\frac{2\sqrt{2}}{3}a^3$; where $2a$=edge

Height=$2a\sqrt{\frac{2}{3}}$; whole surface=$4a^2\sqrt{3}$

31. Wedges on rectangular bases :

Volume=$\frac{bh}{6}(2l+a)$ where h=height,

b=width of base,

$=\frac{A}{3}(2l+a)$ l=length of base, and

a=edge.

A=area of cross-section

32. Wedges on Trapezoidal bases :

Volume$=A.\frac{l_1+l_2+l_3}{3}$ where l_1, l_2, l_3 are the lengths of three parallel edges, and A=ara of cross-section.

33. Oblique frusta of right of circular cylinders :

Volume=transverse sectional area$\times$mean length.

Curved surface=Circumference at transverse section$\times$mean length of frusta.

34. Prismoids :

$$\text{Volume} = \frac{\text{height}}{6}\left\{\text{top area}+\text{base area}+4\ (\text{area at mid-section})\right\}$$

35. Frusta of pyramids and cones :

Volume $=\frac{h}{3}\ (A_1+A_2+\sqrt{A_1A_2})$; where, h=height,

A_1&A_2 are the areas of the end.

36. Frusta of right regular pyramids :

$$\text{Volume} = \frac{h}{3}\left(A_1+A_2+\sqrt{A_1A_2}\right)$$

37. Frusta of right circular cones :

$$\text{Volume} = \frac{\pi h}{3}(R^2+r^2+Rr)$$

curved surface$=\pi(R+r)\times$slant height

where, h=height of the frusta, and

R, r=radii of the ends.

38. Spheres :

Volume $=\frac{\pi d^9}{6}=\frac{4}{3}\pi r^3$ where r=radius, and

whole surface$=\pi d^2=4\pi r^2$ d=diameter.

39. Spherical shells :

Volume $=\frac{\pi}{6}\ (D^3-d^3)$ where d=inner diameter

$$=\frac{4\pi}{3}(R^3=r^3)$$

D=outer diameter

r=inner radius,

and R=outer radius,

40. Oblate spheroids :

Volume $=\frac{4}{3}\pi a^2 b$; where a=semi major axis,

and b=semi minor axis.

41. Prolate spheroids :

Volume $=\frac{4}{3}\pi ab^2$.

42. Segments of Spheres :

Volume $=\frac{\pi h^2}{6}(3d-2h)$

$=\frac{\pi h}{6}(3r^2+h^2)$

where, r=radius of the base of the segment,

h=height, and

d=diameter of the sphere

curved surface$=\pi dh$.

43. Sectors of spheres :

Volume $=\frac{2}{3}\pi r^2 h$

whole surface$=\pi r\left(2h+\sqrt{(2rh-h^2)}\right)$

where, r=radius of sphere and h=height.

44. Simpson's rule (for areas)

$$\text{Any irregular area}=\frac{d}{3}\Big\{(o_1+o_n)+4(o_2+o_4+o_6+\ldots\ldots)+2(o_3+o_5+o_7+\ldots\ldots\ldots\ldots)\Big\}$$

where d=Common distance (between two adjacent ordinates)

o_1=first ordinate ;

o_n=last ordinate ;

$o_2, o_4, o_6......$=even ordinates

$o_3, o_5, o_7......$=odd ordinates

*No. of ordinates must be odd.

45. **Weddle's rule :** (No. of divisions must be a multiple of six)

$$\text{Area}=\frac{3d}{10}\Big\{5(\text{sum of even ordinates})+\text{middle ordinate}+\text{all odd ordinates}\Big\}$$

46. **Trapezoidal rule** (for calculating area) :

$$\text{Area}=d\left(\frac{o_1}{2}+o_2+o_3.........+o_{n-1}+\frac{o_n}{2}\right)$$

47. **Simpson's rule** (for volumes) :

$$\text{Volume}=\frac{L}{3}\left(A_1+AA_2+2A_3+4A_4+......2A_{n-2}+4A_{n-1}+A_n\right)$$

where, L=distance between the sections,

$A_1, A_2, A_3,.........+A_n$=areas of the sections.

48. **Prismoidal formula** (for volumes) :

$$\text{Volume of the prismoid}=\frac{L}{6}(A_1+A_2+4A_m)$$

where, L=length of prismoid.

A_1, A_2=areas of the ends.

A_m=middle area.

49. **Method of end Areas** (for Volumes) :

$$\text{Volume}=L\left(\frac{A_1+A_2}{2}\right)$$

where A_1, A_2, are end areas and L is the length

50. **Middle-Area method** (for volumes) :

Volume=$L\times M$

where=L=Length

M=Mid-area

3

Trigonometry

$180° = \pi$ radians ; $1° = \frac{\pi}{180}$ radians ; 1 radian $= \frac{180}{\pi}$ degrees.

Definition of Functions :

$\text{Sin}\theta = \frac{\text{opposite side}}{\text{hypotenuse}}$, $\cos\theta = \frac{\text{adjacent side}}{\text{hypotenuse}}$;

$\text{ten}\theta = \frac{\text{opposite side}}{\text{adjacent side}}$; $\cot\theta = \frac{\text{adjacent side}}{\text{opposite side}}$;

$\sec\theta = \frac{\text{hypotenuse}}{\text{adjacent side}}$; $\text{cosec}\theta = \frac{\text{hypotenuse}}{\text{opposite side}}$;

$\text{ten}\theta = \frac{\sin\theta}{\cos\theta}$; $\cot\theta = \frac{1}{\tan\theta} = \frac{\cos\theta}{\sin\theta}$

$\sec\theta = \frac{1}{\cos\theta}$; $\text{cosec}\theta = \frac{1}{\sin\theta}$.

Signs of the functions :

Quadrant	Magnitude of angles	sin	cos	tan	cot	sec	cosce
I	0° to 90°	+	+	+	+	+	+
II	90° to 180°	+	−	−	−	−	+
III	180° to 270°	−	−	+	+	−	−
IV	270° to 360°	−	+	−	−	+	−

Values of negative angles :

$\sin(-\theta) = -\sin\theta$ $\qquad$ $\sec(-\theta) = \sec\theta$

$\cos(-\theta) = \cos\theta$ $\qquad$ $\text{cosec}(-\theta) = -\text{cosec}\,\theta$

$\tan(-\theta) = -\tan\theta$ $\qquad$ $\cot(-\theta) = -\cot\theta$

Trigonometric functions of angles (0° to 360°)

Function	$\alpha = 90° \pm \theta$	$\alpha = 180° \pm \theta$	$\alpha = 270° \pm \theta$	$\alpha = 360° \pm \theta$
$\sin\alpha$	$\cos\theta$	$\mp\sin\theta$	$-\cos\theta$	$-\sin\theta$
$\cos\alpha$	$\mp\sin\theta$	$-\cos\theta$	$\pm\sin\theta$	$+\cos\theta$
$\tan\alpha$	$\mp\cot\theta$	$\pm\tan\theta$	$\mp\cot\theta$	$-\tan\theta$
$\cot\alpha$	$\mp\tan\theta$	$\pm\cot\theta$	$\mp\tan\theta$	$-\cot\theta$

Important identities :

$\sin^2\theta + \cos^2\theta = 1 \qquad \therefore \sin\theta = \sqrt{1-\cos^2\theta}$ and $\cos\theta = \sqrt{1-\sin^2\theta}$

$\sec^2\theta = 1 + \tan^2\theta\ ; \qquad \text{cosec}^2\,\theta = 1 + \cot^2\theta$

Relation between the functions of the same angle :

Function	Sin α	Cos α	Tan α	Ctn α	Sec α	Csc α
Sin α	Sin α	$\pm\sqrt{1-\cos^2\alpha}$	$\frac{\tan\alpha}{\pm\sqrt{1+\tan^2\alpha}}$	$\frac{1}{\pm\sqrt{1+\text{ctn}^2\alpha}}$	$\frac{\pm\sqrt{\sec^2\alpha-1}}{\sec\alpha}$	$\frac{1}{\csc\alpha}$
Cos α	$\pm\sqrt{1-\sin^2\alpha}$	$\cos\alpha$	$\frac{1}{\pm\sqrt{1+\tan^2\alpha}}$	$\frac{\text{ctn}\,\alpha}{\pm\sqrt{1+\text{ctn}^2\alpha}}$	$\frac{1}{\sec\alpha}$	$\frac{\pm\sqrt{\csc^2\alpha-1}}{\csc\alpha}$
Tan α	$\frac{\sin\alpha}{\pm\sqrt{1-\sin^2\alpha}}$	$\frac{\pm\sqrt{1-\cos^2\alpha}}{\cos\alpha}$	$\tan\alpha$	$\frac{1}{\text{ctn}\,\alpha}$	$\pm\sqrt{\sec^2\alpha-1}$	$\frac{1}{\pm\sqrt{\csc^2\alpha-1}}$
Ctn α	$\frac{\pm\sqrt{1-\sin^2\alpha}}{\sin\alpha}$	$\frac{\cos\alpha}{\pm\sqrt{1-\cos^2\alpha}}$	$\frac{1}{\tan\alpha}$	$\text{ctn}\,\alpha$	$\frac{1}{\pm\sqrt{\sec^2\alpha-1}}$	$\pm\sqrt{\csc^2\alpha-1}$
Sec α	$\frac{1}{\pm\sqrt{1-\sin^2\alpha}}$	$\frac{1}{\cos\alpha}$	$\pm\sqrt{1+\tan^2\alpha}$	$\frac{\pm\sqrt{1+\text{ctn}^2\alpha}}{\text{ctn}\,\alpha}$	$\sec\alpha$	$\frac{\csc\alpha}{\pm\sqrt{\csc^2\alpha-1}}$
Csc α	$\frac{1}{\sin\alpha}$	$\frac{1}{\pm\sqrt{1-\cos^2\alpha}}$	$\frac{\pm\sqrt{1+\tan^2\alpha}}{\tan\alpha}$	$\pm\sqrt{1+\text{ctn}^2\alpha}$	$\frac{\sec\alpha}{\pm\sqrt{\sec^2\alpha-1}}$	$\csc\alpha$

Summation and Difference of two angles :

$$\sin(\alpha \pm \beta) = \sin \alpha \cos \beta \pm \cos \alpha \sin \beta ;$$

$$\cos(\alpha \pm \beta) = \cos \alpha \cos \beta \mp \sin \alpha \sin \beta ;$$

$$\tan(\alpha \pm \beta) = \frac{\tan \alpha \pm \tan \beta}{1 \mp \tan \alpha \tan \beta} ;$$

$$\cot(\alpha \pm \beta) = \frac{\cot \alpha \cot \beta \mp 1}{\cot \alpha \pm \cot \beta} ;$$

Summation of three angles :

$$\sin(\alpha+\beta+\gamma) = \sin \alpha \cos \beta \cos \gamma - \cos \alpha \sin \beta \cos \gamma + \cos \alpha \cos \beta \sin \gamma - \sin \alpha \sin \beta \sin \gamma$$

$$\cos(\alpha+\beta+\gamma) = \cos \alpha \cos \beta \cos \alpha - \sin \alpha \sin \beta \cos \gamma - \sin \alpha \cos \beta \sin \gamma - \cos \alpha \sin \beta \sin \gamma .$$

Functions of multiple angles :

$$\sin 2\alpha = 2 \sin \alpha \cos \alpha ;$$

$$\cos 2\alpha = \cos^2 \alpha - \sin^2 \alpha = 1 - 2 \sin^2 \alpha = 2 \cos^2 \alpha - 1 ;$$

$$1 - \cos 2\alpha = 2 \sin^2 \alpha ; \qquad 1 + \cos^2 2\alpha = 2 \cos^2 \alpha ;$$

$$\tan^2 \alpha = \frac{1 - \cos 2\alpha}{1 + \cos 2\alpha} ; \qquad \tan 2\alpha = \frac{2 \tan \alpha}{1 - \tan^2 \alpha} ;$$

$$\cot 2\alpha = \frac{\cot^2 \alpha - 1}{2 \cot \alpha} ; \qquad \tan 3\alpha = \frac{3 \tan \alpha - \tan^3 \alpha}{1 - 3\tan^3 \alpha} ;$$

$$\sin 3\alpha = 3 \sin \alpha - 4 \sin^3 \alpha ;$$

$$\cos 3\alpha = 4 \cos^3 \alpha - 3 \cos \alpha ; \quad \cot 3\alpha = \frac{\cot^3 \alpha - 3 \cot \alpha}{3 \cot^2 \alpha - 1} ;$$

$$\sin 4\alpha = 8 \cos^3 \alpha \sin \alpha - 4 \cos \alpha \sin \alpha ;$$

$$\cos 4\alpha = 8 \cos^4 \alpha - 8 \cos^2 \alpha + 1 ;$$

$$\tan 4\alpha = \frac{4 \tan \alpha - 4 \tan^3 \alpha}{1 - 6 \tan^2 \alpha + \tan^4 \alpha} ;$$

$$\cot 4\alpha = \frac{\cot^4 \alpha - 6 \cot^2 \alpha + 1}{4 \cot^3 \alpha - 4 \cot \alpha} ;$$

Also, $\sin 2\alpha = \dfrac{2 \tan \alpha}{1 + \tan^2 \alpha}$ and $\cos 2\alpha = \dfrac{1 - \tan^2 \alpha}{1 + \tan^2 \alpha}$

Functions of sub-multiple angles

$$\sin\alpha = 2\sin\frac{\alpha}{2}\cos\frac{\alpha}{2}\,; \qquad \tan\alpha = \frac{2\tan\frac{\alpha}{2}}{1-\tan^2\frac{\alpha}{2}}\,;$$

$$\cos\alpha = \cos^2\frac{\alpha}{2} - \sin^2\frac{\alpha}{2}\,; \qquad \cot\alpha = \frac{\cot^2\frac{\alpha}{2}-1}{2\cot\frac{\alpha}{2}}\,;$$

$$= 2\cos^2\frac{a}{2} - 1\,;$$

$$= 1 - 2\sin^2\frac{a}{2}\,; \qquad 2\cos^2\frac{a}{2} = 1 + \cos a\,;$$

$$\tan^2\frac{a}{2} = \frac{1-\cos\alpha}{1+\cos a}\,; \qquad 2\sin^2\frac{a}{2} = 1 - \cos a\,;$$

$$\sin\frac{a}{2} = \sqrt{\frac{1}{2}(1-\cos a)}\,; \qquad \tan\frac{a}{2} = \sqrt{\frac{1-\cos a}{1+\cos a}}\,;$$

$$\cos\frac{a}{2} = \sqrt{\frac{1}{2}(1+\cos a)}\,; \qquad \cot\frac{a}{2} = \sqrt{\frac{1+\cos a}{1-\cos a}}\,;$$

$$\sin a = \frac{2\tan\frac{a}{2}}{1+\tan^2\frac{a}{2}}\,; \qquad \tan a = \frac{2\tan\frac{a}{2}}{1-\tan^2\frac{a}{2}}$$

$$\cos a = \frac{1-\tan^2\frac{a}{2}}{1+\tan^2\frac{a}{2}}\,; \qquad \cot a = \frac{\cot^2\frac{a}{2}-1}{2\cot\frac{a}{2}}\,;$$

Sum and difference of two functions:

$$\sin a + \sin\beta = 2\sin\frac{a+\beta}{2}\cos\frac{a-\beta}{2}\,;$$

$$\sin a - \sin\beta = 2\cos\frac{a+\beta}{2}\sin\frac{a-\beta}{2}\,;$$

$$\cos a + \cos\beta = 2\cos\frac{a+\beta}{2}\cos\frac{a-\beta}{2}\,;$$

$$\cos \alpha - \cos\beta = -2 \sin \frac{\alpha+\beta}{2} \sin \frac{\alpha+\beta}{2};$$

$$\tan \alpha \pm \tan \beta = \frac{\sin(\alpha\pm\beta)}{\cos \alpha \cos\beta}; \quad \cot \alpha \pm \cot \beta = \pm\frac{\sin(\alpha\pm\beta)}{\sin \alpha \sin\beta}$$

$$\tan \alpha + \cot \beta = \frac{\cos(\alpha-\beta)}{\cos \alpha \sin\beta}; \quad \cot \alpha - \tan \beta = \frac{\cos(\alpha+\beta)}{\sin \alpha \cos \beta}.$$

Product of two functions :

$$\sin \alpha \sin \beta = \frac{1}{2} \{\cos(\alpha-\beta) - \cos(\alpha+\beta)\}.$$

$$\cos \alpha \cos \beta = \frac{1}{2} \{\cos(\alpha-\beta) + \cos(\alpha+\beta)\}.$$

$$\sin \alpha \cos \beta = \frac{1}{2} \{\sin(\alpha-\beta) + \sin(\alpha+\beta)\}.$$

Power of functions :

$$\sin^2 \alpha = \frac{1}{2}(1-\cos 2\alpha); \quad \sin^3\alpha = \frac{1}{4}(3 \sin o - \sin 3\alpha);$$

$$\cos^2 \alpha = \frac{1}{2}(1+\cos 2\alpha); \quad \cos^3\alpha = \frac{1}{4}(\cos 3\alpha + 3 \cos \alpha);$$

Inverse functions (Anti-trigonometric functions) **:**

By definition, $x=\sin y$, then $y=\sin^{-1}x$.

y is measured in radians.

Relations among inverse functions :

$$\sin^{-1} a = \cos^{-1}\sqrt{1-a^2} = \tan^{-1}\frac{a}{\sqrt{1-a^2}} = \cot^{-1}\frac{\sqrt{1-a^2}}{a}$$

$$= \sec^{-1}\frac{1}{\sqrt{1-a^2}} = \sec^{-1}\frac{1}{a}.$$

$$\cos^{-1} a = \sin^{-1}\sqrt{1-a^2} = \tan^{-1}\frac{\sqrt{1-a^2}}{a} = \cot^{-1}\frac{a}{\sqrt{1-a^2}}$$

$$= \sec^{-1}\frac{1}{a} = \cos^{-1}\frac{1}{\sqrt{1-a^2}}.$$

$$\tan^{-1} a = \sin^{-1}\frac{a}{\sqrt{1+a^2}} = \cos^{-1}\frac{1}{\sqrt{1+a^2}} = \cot^{-1}\frac{1}{a}$$

$$=\sec^{-1}\sqrt{1+a^2}=\cos^{-1}\frac{\sqrt{1+a^2}}{a}.$$

$$\cot^{-1}a=\tan^{-1}\frac{1}{a};\ \sec^{-1}a=\cos^{-1}\frac{1}{a};\ \cos^{-1}a=\sin^{-1}\frac{1}{a}.$$

Important relations between inverse functions :

$\text{Sin}^{-1}x+\sin^{-1}y=\sin^{-1}(x\sqrt{1-y^2}+y\sqrt{1-x^2})$ when $xy\leqslant 0$

$\text{Sin}^{-1}x-\sin^{-1}y=\sin^{-1}(x\sqrt{1-y^2}-y\sqrt{1-x^2})$ when $xy\geqslant 0$

$\text{Cos}^{-}x+\cos^{-1}y=\cos^{-1}(xy-\sqrt{1x^2}\sqrt{1-y^2})$ when $x+y\geqslant 0$

$\text{Cos}^{-1}x-\cos^{-1}y=\cos^{-1}(xy+\sqrt{1-x^2}.\ \sqrt{1-y^2})$ when $x<y$

$\tan^{-1}x+\tan^{-1}y=\tan^{-1}\dfrac{x+y}{1-xy}$ when $xy<1$

$\tan^{-1}x-\tan^{-1}y=\tan^{-1}\dfrac{x-y}{1+xy}$ when $xy>-1$

$2\ \text{Sin}^{-1}x=\text{Sin}^{-1}(2x\sqrt{1-x^2})$ when $|x|\leqslant\dfrac{1}{\sqrt{2}}$

$2\ \text{Cos}^{-1}x=\cos^{-1}(2x^2-1)$ when $0\leqslant x\leqslant 1$

$2\tan^{-1}x=\tan^{-1}\dfrac{2x}{1-x^2}$ when1 $|x|<1$

Solution of triangles :

Right triangles : $\text{Sin}\ \alpha=\dfrac{a}{c}$, $\cos\alpha=\dfrac{b}{c}$, $\tan\alpha=\dfrac{a}{b}$

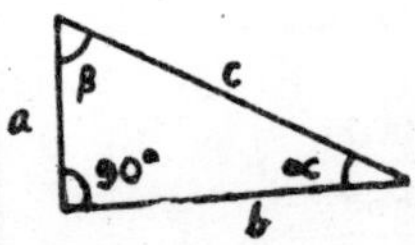

$$a=c\ \text{Sin}\ \alpha=b\tan\alpha=\sqrt{c^2-b^2}.$$

$$b=c\cos a=\frac{a}{\tan\alpha}=\sqrt{c^2-b^2}$$

$$c=\frac{a}{\sin\alpha}=\frac{b}{\cos\alpha}=\sqrt{a^2+b^2}.$$

$$\text{Area}=\frac{1}{2}ab=\frac{a^2}{2\tan\alpha}=\frac{b^2\tan\alpha}{2}=\frac{c^2\tan 2\alpha}{4}$$

Oblique triangles :

Angle α oppostte to side a
Angle β opposite to side b
Angle γ opposite to side c

$$\frac{a}{\sin \alpha}=\frac{b}{\sin \beta}=\frac{c}{\sin \gamma}=2\text{ R}$$ where R=radius of circumseribed circle.

$$\frac{a+b}{a-b}=\frac{\tan \frac{\alpha+\beta}{2}}{\tan \frac{\alpha-\beta}{2}}$$

$$\text{Area}=\frac{1}{2}\,ab \sin \gamma = 2R^2 \sin \alpha \sin \beta \sin \gamma$$

$$=\sqrt{s(s-a)(s-b)(s-c)}.$$

$$a=b \cos \gamma+c \cos \beta$$

$$a^2=b^2+c^2-2bc \cos \alpha$$

Given datas	Required Solutions
Sides a, b, c	$r=\sqrt{\frac{(s-a)(s-b)(s-c)}{s}}$; $A=\sqrt{s(s-a)(s-b)(s-c)}$
one side, a and two angles α, β	$A=\frac{1}{2}\,ac \sin \beta$; $b=a.\frac{\sin \beta}{\sin \alpha}$ and $c=a.\frac{\sin \gamma}{\sin \alpha}$.
Two sides a, b, and included angle γ	$A=\frac{1}{2}\,ab \sin \gamma$; $c=a.\frac{\sin \gamma}{\sin \alpha}$.
Two sides, a, b, and one of the opposite angles, α	$\sin \beta=\frac{b \sin \alpha}{a}$; $c=a\frac{\sin \gamma}{\sin \alpha}$. $A=\frac{1}{2}\,ab \sin \gamma$.

Relations derived from the above relations :

$$\tan \alpha=\frac{a \sin \beta}{c-a \cos \beta}\ ;\ \tan\frac{\alpha}{2}=\sqrt{\frac{(s-b)(s-c)}{s(s-a)}}.$$

$$\frac{\propto}{2}=\sqrt{\frac{(s-b)(s-e)}{ab}}\;;\;\cos=\frac{\propto}{2}=\sqrt{\frac{s(s-a)}{bc}}$$

$$=\sqrt{\frac{(s-a)(s-b)(s-c)}{S}}=s\tan\frac{\propto}{2}\tan\frac{\beta}{9}\tan\frac{\gamma}{2}$$

$$=4R\sin\frac{\propto}{2}\sin\frac{\beta}{2}\sin\frac{\gamma}{2}$$

where α=radius of the inscribed circle,
S=semi perimeter of the triangle.

Napier's rules of circular parts :

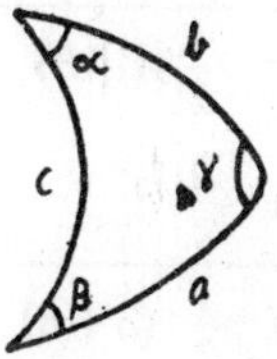

Product of the tangents of the adjacent parts=product of the cosines of the opposite parts.

Right spherical triangle :

$$\sin a=\sin\alpha\sin c=\tan b\cot\beta$$
$$\sin b=\sin\beta\sin c=\tan a\cot\alpha$$
$$\cos c=\cos a\cos b=\cos\alpha\cot\beta$$
$$\cos\alpha=\cos a\sin\beta=\tan b\cot c$$
$$\cos\beta=\cos b\sin\alpha=\tan a\cot c$$

Oblique spherical triangle :

$$\frac{\sin a}{\sin\alpha}=\frac{\sin b}{\sin\beta}=\frac{\sin c}{\sin\gamma}\;;\qquad s=\frac{a+b+c}{2}$$

$$\sigma=\frac{\alpha+\beta+\gamma}{2}$$

$$\cos a=\cos b\cos c+\sin b\sin c\cos\alpha\;;$$
$$\cos\alpha=-\cos\beta\cos\gamma+\sin\beta\sin\gamma\cos a\;;$$

$$\tan\frac{a}{2}=\sqrt{\frac{-\cos\sigma.\cos(\sigma-\alpha)}{\cos(\sigma-\beta).\cos(\sigma-\gamma)}}$$

$$\tan\frac{\alpha}{2}=\sqrt{\frac{\sin(s-b)\sin(s-c)}{\sin s.\sin(s-a)}}$$

$$\sin\frac{1}{2}(\alpha+\beta)\cos\frac{1}{2}c=\cos\frac{1}{2}(a-b)\cos\frac{\gamma}{2}$$

$$\cos\frac{1}{2}(\alpha+\beta)\cos\frac{1}{2}c=\cos\frac{1}{2}(a+b)\sin\frac{\gamma}{2}$$

$$\sin\frac{1}{2}(\alpha-\beta)\sin\frac{1}{2}c=\sin\frac{1}{2}(a-b)\cos\frac{\gamma}{2}$$

$$\cos\frac{1}{2}(\alpha-\beta)\sin\frac{1}{2}c=\sin\frac{1}{2}(a+b)\sin\frac{\gamma}{2}$$

$$\frac{\sin\frac{1}{2}(a+b)}{\sin\frac{1}{2}(a-b)}=\frac{\cot\frac{\gamma}{2}}{\tan\frac{1}{2}(\alpha-\beta)};\quad \frac{\cos\frac{1}{2}(a+b)}{\cos\frac{1}{2}(a-b)}=\frac{\cot\frac{\gamma}{2}}{\tan\frac{1}{2}(\alpha+\beta)}$$

$$\frac{\sin\frac{1}{2}(\alpha+\beta)}{\sin\frac{1}{2}(\alpha-\beta)}=\frac{\tan\frac{1}{2}c}{\tan\frac{1}{2}(a-b)};\quad \frac{\cos\frac{1}{2}(\alpha+\beta)}{\cos\frac{1}{2}(\alpha-\beta)}=\frac{\tan\frac{1}{2}c}{\tan\frac{1}{2}(a+b)}$$

$\text{Area}=\dfrac{\pi R^2 E}{180}$; Where R=radius of the sphere upon which the triangle lies,

and E=spherical excess

$=\alpha+\beta+\gamma-180°$

Hyperbolic functions :

$$\sinh x=\frac{e^x-e^{-x}}{2};\quad \cosh x=\frac{e^x+e^{-x}}{2};\quad \tanh x=\frac{e^x-e^{-x}}{e^x+e^{-x}}$$

$$\coth x=\frac{e^x+e^{-x}}{e^x-e^{-x}};\quad \text{seeh}\ x\ \frac{2}{e^{-x}+e^{-x}};\quad \text{cosech}\ x=\frac{2}{e^x e^{-x}}$$

Anti-hyperbolic functions :

$$\sinh^{-1}x=\log_e\left(x+\sqrt{x^2+1}\right);\quad \cosh^{-1}x=\log_e\left(x+\sqrt{x^2-1}\right);$$

$$\tanh^{-1}x=\tfrac{1}{2}\log_e\left(\frac{1+x}{1-x}\right);\quad \coth^{-1}x=\tfrac{1}{2}\log_e\left(\frac{x+1}{x-1}\right)$$

$$\text{sech}^{-1}x=\log_e\left(\frac{1+\sqrt{1-x^2}}{x}\right);\quad \text{cosech}^{-1}x=\log_e\left(\frac{1+\sqrt{1+x^2}}{x}\right)$$

Identities related to hyperbolic functions :

$\cosh^2 x - \sin h^2 x = 1$; $\sinh 2x = 2 \sinh x \cosh x$;

$\operatorname{sech}^2 x + \tan h^2 x = 1$; $\cosh 2x = \cosh^2 x + \sin h^2 x$;

$\coth^2 x - \operatorname{cosech}^2 x = 1$;

$2 \sinh^2 \frac{x}{2} = \cosh x - 1$; $\tanh (x \pm y) = \frac{\tanh x \pm \tanh y}{1 \pm \tanh x \tanh y}$;

$2 \cosh^2 \frac{x}{2} = \cosh x + 1$;

$\sinh (x \pm y) = \sinh x \cosh y \pm \cos h\, x \sinh y$;

$\cosh (x \pm y) = \cosh x \cosh y \pm \sin h\, x \sinh y$;

Relation with circular functions ;

$\sin x = \frac{e^{ix} - e^{-ix}}{2i}$; $\cos x = \frac{e^{ix} + e^{-ix}}{2} = \cosh ix$;

$= -i \sinh ix$: $\tan x = -i \tanh ix$;

$\sin ix = i \sinh x$ | $\sinh ix = i \sin x$

$\cos ix = \cosh x$ | $\cosh ix = \cos x$

$\tan ix = i \tanh x$ | $\tanh ix = i \tan x$

$e\theta = \cosh\theta + \sinh \theta$ | $e^{i}\theta = \cos \theta + i \sin \theta$

$e^{-}\theta = \cosh\theta - \sinh \theta$ | $e^{-i}\theta = \cos \theta - i \sin \theta$

$\sinh (x \pm iy) = \sinh x \cos y \pm i \cosh x \sin y$

$\cosh (x \pm iy) = \cos h\, x \cos y \pm i \sinh x \sin y$

$$\sinh x + \sinh y = 2 \sinh \left(\frac{x+y}{2}\right) \cos h \left(\frac{x-y}{2}\right)$$

$$\sinh x - \sinh y = 2 \cosh \left(\frac{x+y}{2}\right) \sinh \left(\frac{x-y}{2}\right)$$

$$\cosh x + \cosh y = 2 \cosh \left(\frac{x+y}{2}\right) \cosh \left(\frac{x-y}{2}\right)$$

$$\cosh x - \cosh y = 2 \sinh \left(\frac{x+y}{2}\right) \sinh \left(\frac{x-y}{2}\right)$$

4

Analytic Geometry (Plane and Solid)

Relation between rectangular and polar coordinates :

$x = r\cos\theta$
$y = r\sin\theta$

$r = \sqrt{x^2+y^2}$
$\theta = \tan^{-1}\frac{y}{x}$

$\sin\theta = \frac{y}{\sqrt{x^2+y^2}}$
$\cos\theta = \frac{x}{\sqrt{x^2+y^2}}$

Points and Slopes :

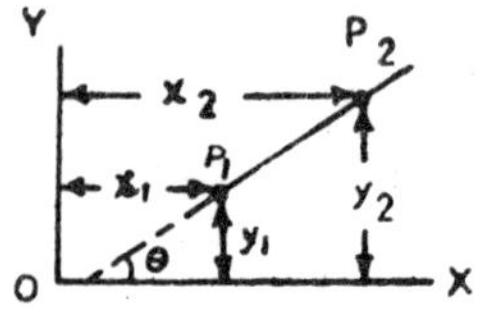

Distance between P_1 and P_2
$= \sqrt{(x_2-x_1)^2+(y_2-y_1)^2}$

Slope of $P_1P_2 = \tan\theta = m = \frac{y_2-y_1}{x_2-x_1}$

Mid-point of $P_1P_2 \rightarrow \left(\frac{x_1+x_2}{2}, \frac{y_1+y_2}{2}\right)$

Point dividing P_1P_2 in ratio $m_1 : m_2 \rightarrow \left(\frac{m_1x_2+m_2x_1}{m_1+m_2}, \frac{m_1y_2+m_2y_1}{m_1+m_2}\right)$

Straight Line :

(1) $y = mx+c$.........where, m=slope, c=intercept on y-axis

(2) $\frac{x}{a}+\frac{y}{b}=1$......where, a and b are the intercepts on x and y-axes respectively.

(3) $x\cos\theta+y\sin\theta=P$.........where P=distance from origin to the line.

(4) $Ax+By+C=0$.........general form.

(5) $y-y_1=m(x-x_1)$.........when line passes through point (x_1, y_1)

(6) $\frac{y-y_1}{x-x_1}=\frac{y_2-y_1}{x_2-x_1}$.........when line passes through points (x_1, y_1) and (x_2, y_2).

Circle :

$(x-l)^2+(y-m)^2=R^2$.........general eqn ;
where l and m are coordinates at centre
and R=radius of the circle.

Parabola :

When l and m are coordinates at vertex of the parabola,

$$(y-m)^2=4a(x-l) \quad ...(1)$$

$$(x-l)^2=4a(y-m) \quad ...(2)$$

a=distance from vertex to focus
=distance from vertax to directrix
$4a$=latus rectum.

Ellipse :

Semi-major axis$=a$, Semi-minor axis$=b$.

Distance from centre to either focus$=\sqrt{a^2-b^2}$

Distance from centre to either directrix$=\dfrac{a}{\text{eccentricity}}$

Latus rectum$=\dfrac{2b^2}{a}$. Eccentricity$=\dfrac{\sqrt{a^2-b^2}}{a}$

$$\frac{(x-l)^2}{a^2}+\frac{(y-m)^2}{b^2}=1 \quad \text{or,} \quad \frac{(y-m)^2}{a^3}+\frac{(x-l)^2}{b^2}=1$$

Hyperbola :

Transverse axis$=2a$, conjugate axis$=2b$.

Distance from centre to either focus$=\sqrt{a^2+b^2}$

Distance from centre to either directrix$=\dfrac{a}{\text{eccentricity}}$

Eccentricity$=\dfrac{\sqrt{a^2+b^2}}{a}$: Latus rectum$=\dfrac{2b^2}{a}$

$$\frac{(x-l)^2}{a^2}-\frac{(y-m)^2}{b^2}=1 \quad \text{or,} \quad \frac{(y-m)^2}{a^2}-\frac{(x-l)^2}{b^2}=1$$

EQUATIONS OF HIGHER PLANE CURVES :

Sine Curve ;

$y = a \sin cx$;

Area under one arch $= \frac{2a}{c}$;

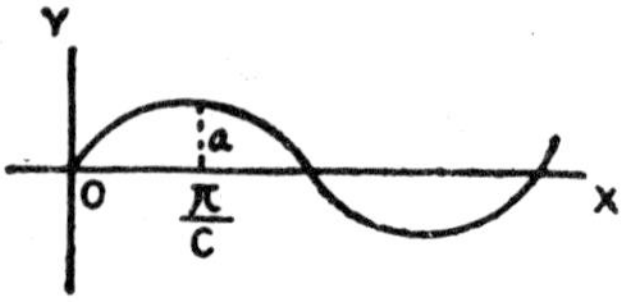

Cosine Curve :

$y = a \cos cx$;

Area under one arch $= \frac{2a}{c}$;

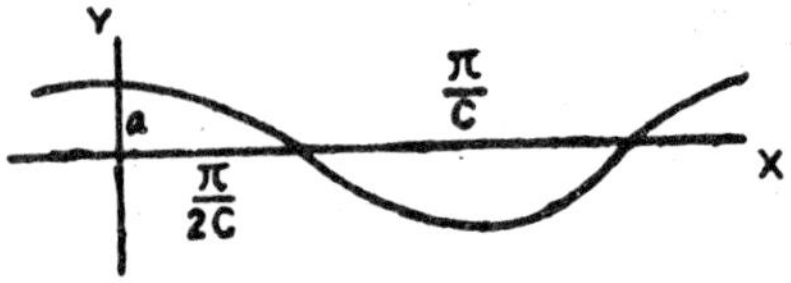

Wave Curve :

$y = a \sin cx + b \cos cx$

Area under one arch

$$= \frac{2\sqrt{a^2+b^2}}{c}$$

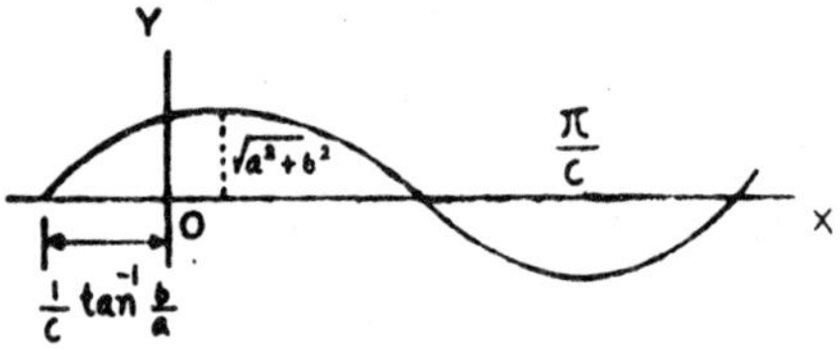

Cardioid :

$x^2+y^2+ax = a\sqrt{x^2+y^2}$

$x = a \cos t\,(1-\cos t)$

$y = a \sin t\,(1-\cos t)$

$r = a\,(1-\cos\theta)$

Area $= \frac{3\pi a^2}{2}$ Length $= 8a$

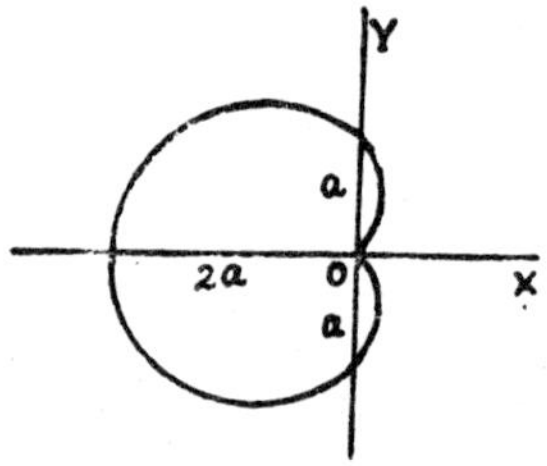

Cycloid :

$x = a\,(t-\sin t)$

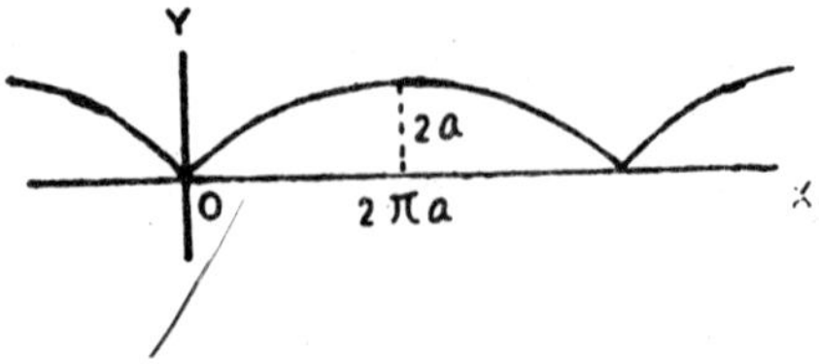

Hypocycloid

$x^{2/3}+y^{2/3}=a^{2/3}$

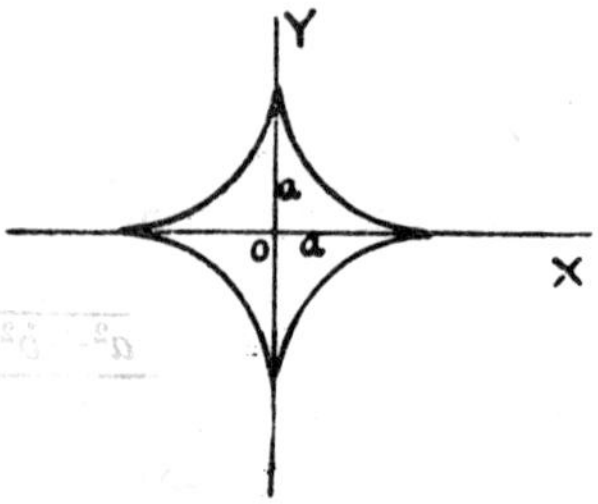

$y=a\ (1-\cos t)$

Area under one arch $=3\pi a^2$

Length of one arch $=8a$.

$x=a\cos^3 t \qquad y=a\sin^3 t$

$\text{Area}=\frac{3\pi a^2}{8} \qquad \text{Length}=6a$

Sine Curve :

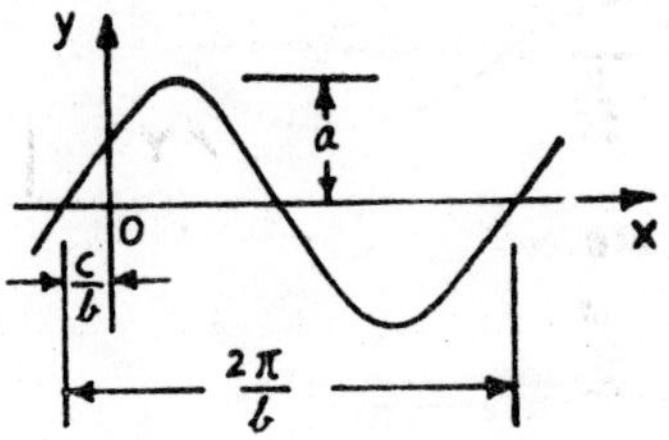

a=amplitude=max. height of wave.

$\frac{2\pi}{b}$ =wave length

x=phase$=-\frac{c}{b}$

$y=a\sin(bx+c)$

Trigonometric Curves :

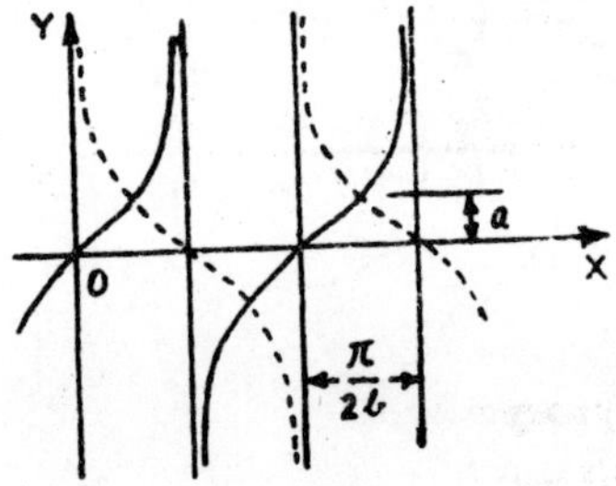

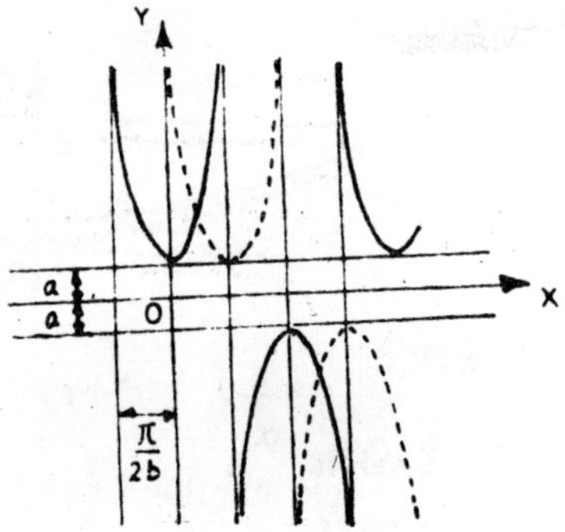

$y=a\tan bx$

$\therefore\quad x=\frac{1}{b}\tan^{-1}\left(\frac{y}{a}\right)$

$y=a\cot bx$

$\therefore\quad x=\frac{1}{b}\cot^{-1}\left(\frac{y}{a}\right)$

$y=a\sec bx=a\ \text{cosec}\ bx$

$\therefore\ x=\frac{1}{b}\sec^{-1}\left(\frac{y}{a}\right)$

$=\frac{1}{b}\text{cosec}^{-1}\left(\frac{y}{a}\right)$

Probability Curve :

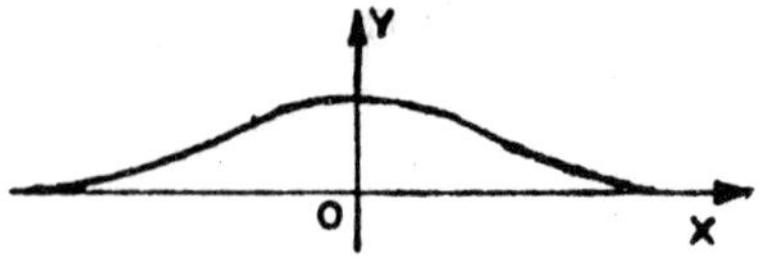

$y=e^{-x^2}$

Oscillatory wave :

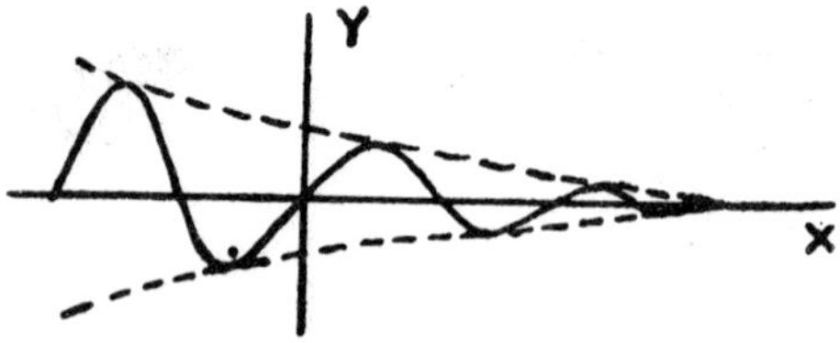

$y=e^{-ax}\sin bx$

Catenary :

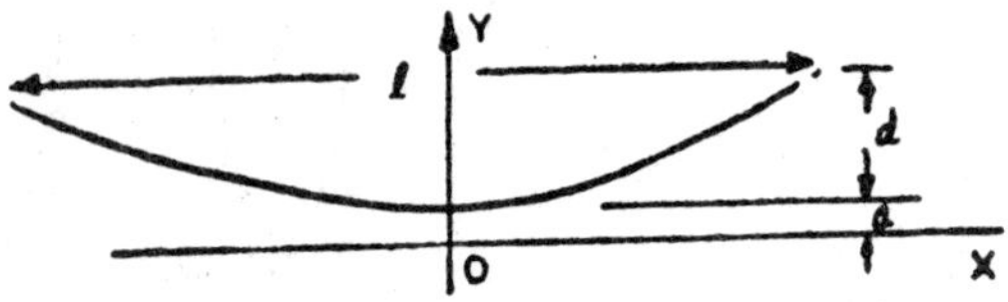

$$y=\frac{a}{2}\left(e^{x/a}+e^{-x/a}\right)$$

$$\text{Arc length}=l\left\{1+\frac{2}{3}\left(\frac{2d}{l}\right)^2\right\}.$$

Logarithmic Curve :

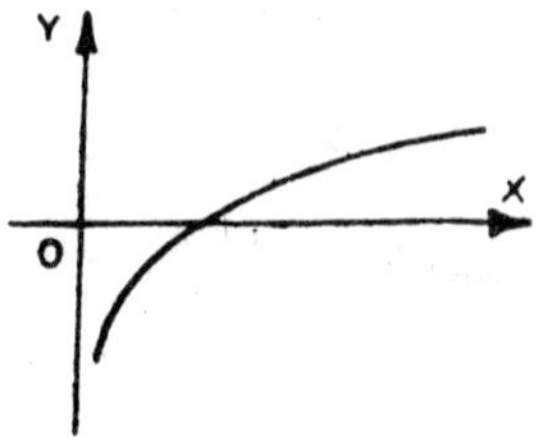

$y=\log_a x$ or, $x=a^y$

Exponential Curve :

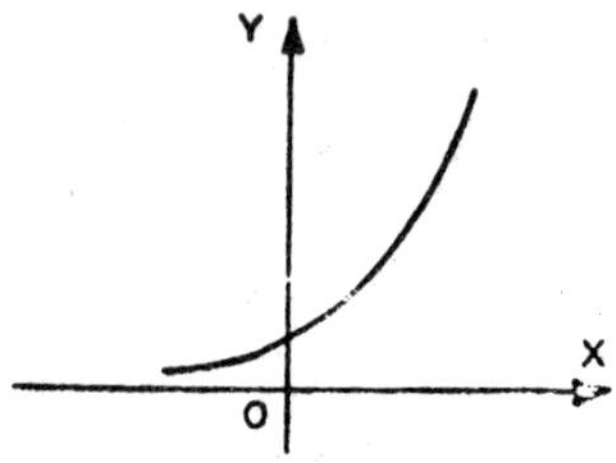

$y=a^x$ or $x=\log_a y$

Cycloid :

$x=a(\phi-\sin\phi)$; $y=a(1-\cos\phi)$;

Area of one arch $=3\pi a^2$;

Length of one earch $=8a$;

Epicycloid :

$$x=(a+b)\cos\phi-a\cos\left(\frac{a+b}{a}\phi\right);$$

$$y=(a+b)\sin\phi-a\sin\left(\frac{a+b}{a}\phi\right)$$

Hypocycloid :

$$x=(a-b)\cos\phi+b\cos\left(\frac{a-b}{b}\phi\right);$$

$$y=(a-b)\sin\phi-b\sin\left(\frac{a-b)}{b}\phi\right)$$

Prolate cycloid :

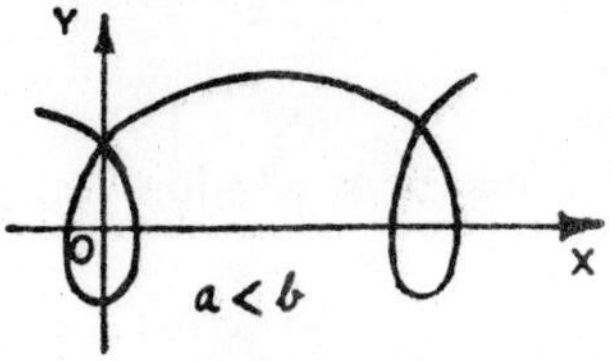

$x=a\phi-b\sin\phi$;

Curtate cycloid

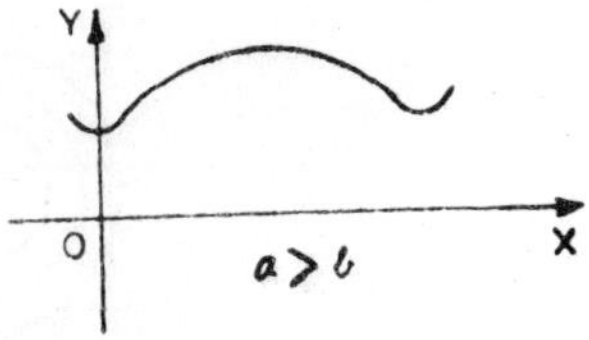

$y=a-b\cos\phi$.

Involute of a circle :

$x=a\cos\phi+a\phi\sin\phi$;

$y=a\sin\phi-a\phi\cos\phi$;

Cardioid:

$r=a(1+\cos\phi)$

Lemniscate :

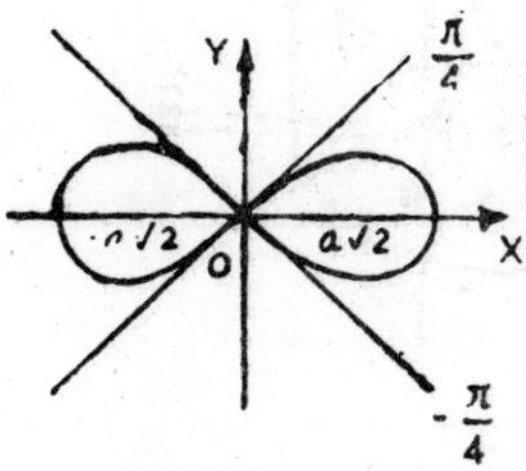

$r^2=2a^2\cos 2\theta$

N-leaved Rose :

$r=a\sin n\theta$

$r=a\cos n\theta$

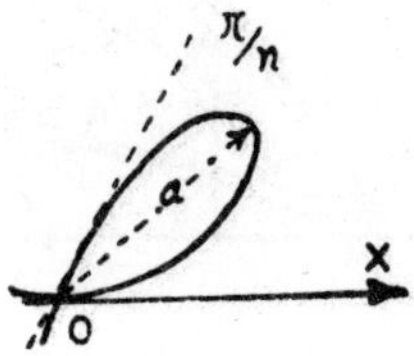

Hyperbolic Spiral :

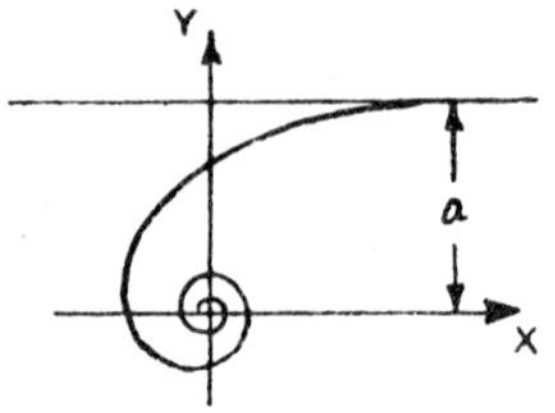

$r\theta=a$

Archimedian Spiral :

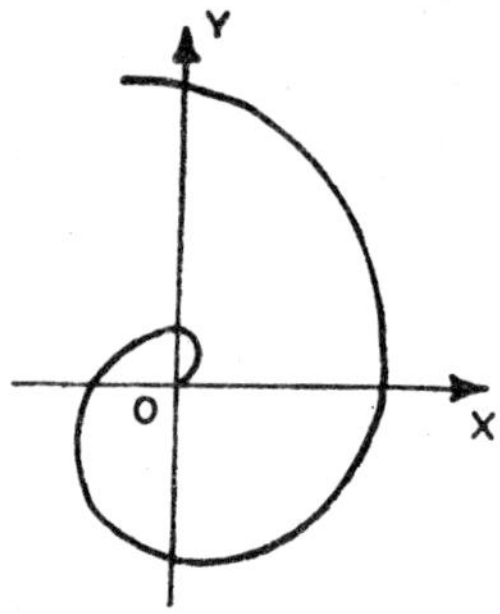

$r=a\ \theta$

Logarithmic Spiral

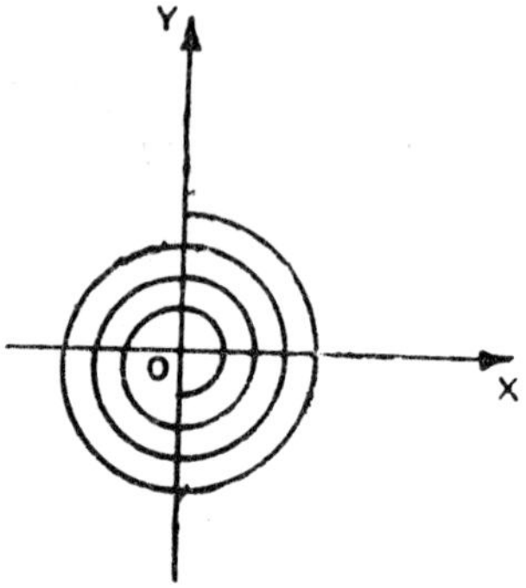

$r=e^{a\theta}$ or $a\theta=\log_e r$

Plane :

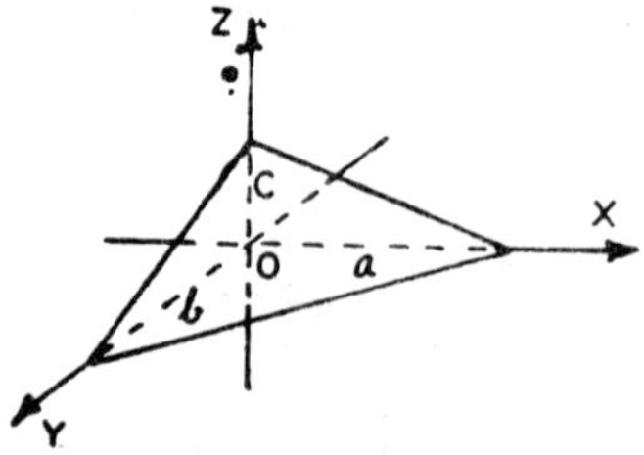

$$\frac{x}{a}+\frac{y}{6}+\frac{z}{c}=1$$

Elliptic Cylinder :

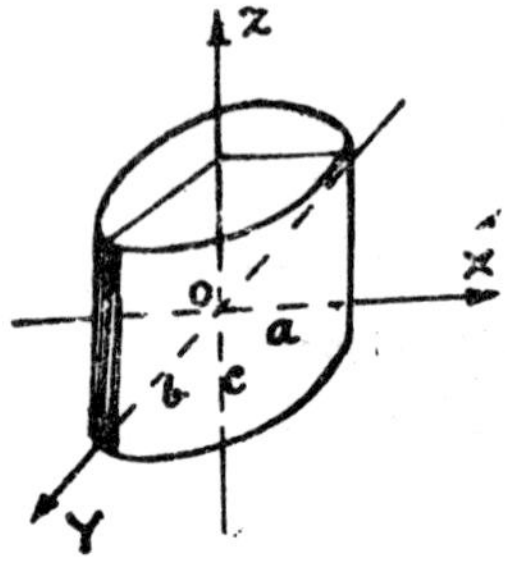

$$\frac{x^2}{a^2}+\frac{y^2}{b^2}=1$$

Sphere :

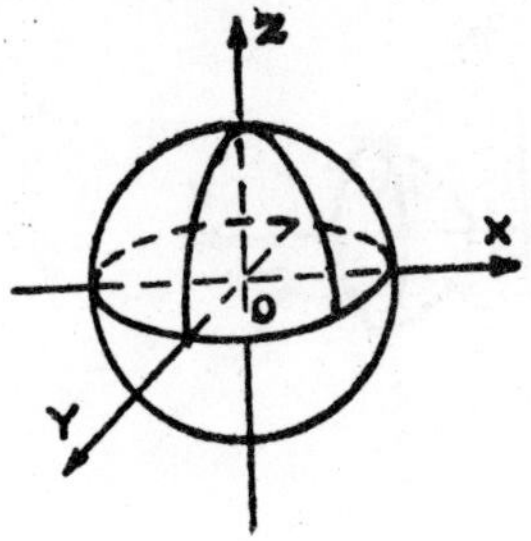

$$x^2+y^2+z^2=r^2$$

Ellipsoid :

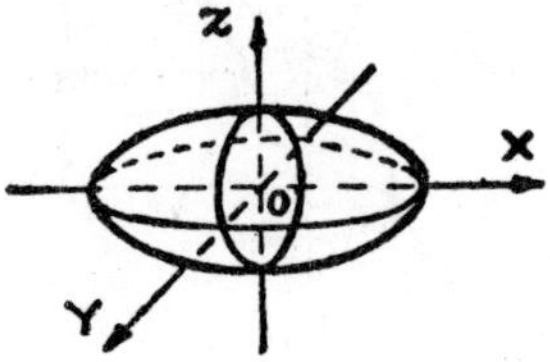

$$\frac{x^2}{a^2}+\frac{y^2}{b^2}+\frac{z^2}{c^2}=1$$

Elliptic Paraboloid :

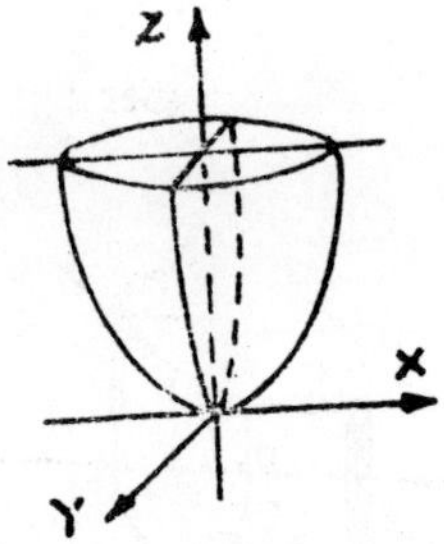

$$\frac{x^2}{a^2}+\frac{y^2}{b^2}=cz$$

Portion of cone

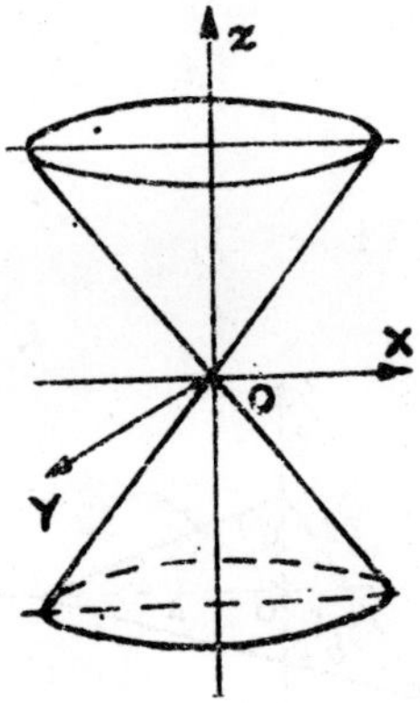

$$\frac{x^2}{a^2}+\frac{y^2}{b^2}-\frac{z^2}{c^2}=o$$

Hyperboloid of one sheet :

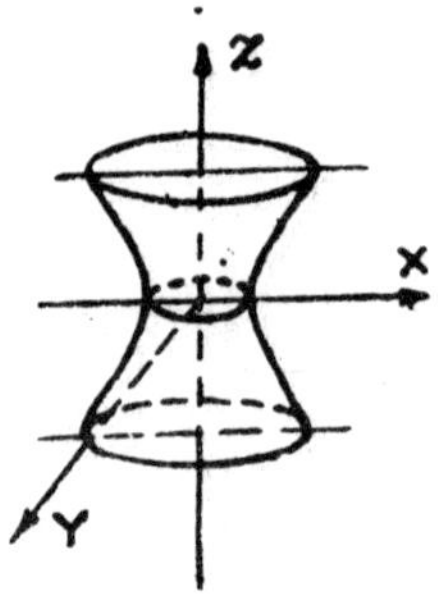

$$\frac{x^2}{a^2}+\frac{y^2}{b^2}-\frac{z^2}{c^2}=1$$

Hyperboloid of two sheets :

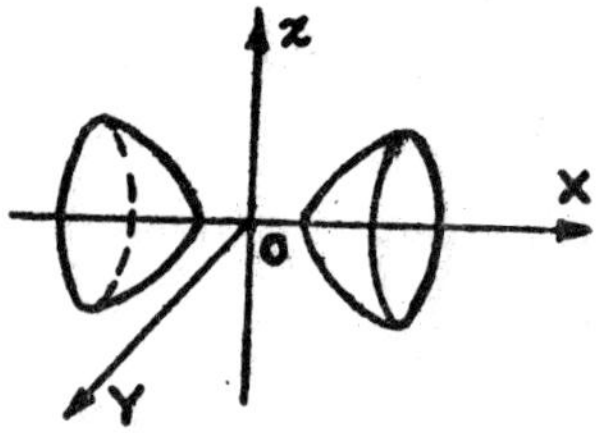

$$\frac{x^2}{a^2}-\frac{y^2}{b^2}-\frac{z^2}{c^2}=1$$

Hyperbolic Paraboloid :

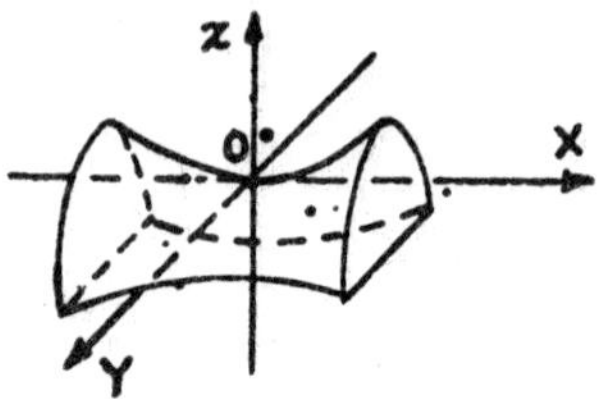

$$\frac{x^2}{a^2}-\frac{y^2}{b^2}=cz$$

5

Differential Calculus

Derivatives :

$\frac{d}{dx}(a)=0 \quad \frac{d}{dx}(x)=1.$

$\frac{d}{dx}(au)=a \frac{du}{dx}$

$\frac{d}{dx}(u \pm v)=\frac{du}{dx} \pm \frac{dv}{dx}.$

$\frac{d}{dx}(uv)=u \frac{dv}{dx}+v \frac{du}{dx}.$

$\frac{d}{dx}\left(\frac{u}{v}\right)=\frac{v \frac{du}{dx} - u \frac{dv}{dx}}{v^2}.$

$\frac{d}{dx}(x^n)= nx^{n-1}.$

$\frac{d}{dx}(a^x)=a^x \log_e a.$

$\frac{d}{dx}(e^x)=e^x.$

$\frac{d}{dx}(\log_e x) = \frac{1}{x}.$

$\frac{d}{dx}(\log_a x)=\frac{\log_a e}{x}.$

$\frac{d}{dx}(u^n)=nu^{n-1}. \frac{du}{dx}.$

$\frac{d}{dx}(\log_a u) = \frac{\log_a e}{u} . \frac{du}{dx}.$

$\frac{d}{dx}(\log_e u)= \frac{1}{u} . \frac{du}{dx}.$

$\frac{d}{dx}(a^u)=a^u. \log_e a. \frac{du}{dx}.$

$\frac{d}{dx}(e^u)=e^u \frac{du}{dx}.$

$\frac{d}{dx}(u^v)=vu^{v-1}\frac{du}{dx}+u^v \log_e u \frac{dv}{dx}.$

$\frac{d}{dx}(\sin x)=\cos x.$

$\frac{d}{dx}(\cos x)=-\sin x.$

$\frac{d}{dx}(\tan x)=\sec^2 x.$

$\frac{d}{dx}(\cot x)=-\text{cosec}^2 x.$

$\frac{d}{dx}(\sec x)= \sec x \tan x.$

$\frac{d}{dx}(\text{cosec } x)=-\text{cosec } x \cot x.$

$\frac{d}{dx}(\sin u)=\cos u \frac{du}{dx}.$

$\frac{d}{dx}(\cos u)=-\sin u \frac{du}{dx}.$

$\frac{d}{dx}(\tan u)=\sec^2 u \frac{du}{dx}.$

$\frac{d}{dx}(\cot u)=-\text{cosec}^2 u \frac{du}{dx}.$

$\frac{d}{dx}(\sec u)=\sec u \tan u \frac{du}{dx}.$

$\frac{d}{dx}(\text{cosce } u)=-\text{cosce } u \cot u \frac{du}{dx}.$

$$\frac{d}{dx}(\text{vers } u) = \sin u \frac{du}{dx}.$$

$$\frac{d}{dx}(\sin^{-1}x) = \frac{1}{\sqrt{1-x^2}}$$

$$\frac{d}{dx}(\cos^{-1}x) = -\frac{1}{\sqrt{1-x^2}}.$$

$$\frac{d}{dx}(\tan^{-1}x) = \frac{1}{1+x^2}.$$

$$\frac{d}{dx}(\cot^{-1}x) = -\frac{1}{1+x^2}.$$

$$\frac{d}{dx}(\sec^{-1}x) = \frac{1}{x\sqrt{x^{2-1}}}.$$

$$\frac{d}{dx}(\text{cosec}^{-1}x) = -\frac{1}{x\sqrt{x^2-1}}.$$

$$\frac{d}{dx}(\sin^{-1}u) = \frac{1}{\sqrt{1-u^2}} \cdot \frac{du}{dx}.$$

$$\frac{d}{dx}(\cos^{-1}u) = -\frac{1}{\sqrt{1-u^2}} \frac{du}{dx}.$$

$$\frac{d}{dx}(\tan^{-1}u) = \frac{1}{1+u^2} \cdot \frac{du}{dx}.$$

$$\frac{d}{dx}(\sec^{-1}u) = \frac{1}{u\sqrt{u^2-1}} \cdot \frac{du}{dx}.$$

$$\frac{d}{dx}\text{vers}^{-1}u) = \frac{1}{\sqrt{2u-u^2}} \cdot \frac{du}{dx}.$$

$$\frac{d}{dx}(\sinh x) = \cosh x.$$

$$\frac{d}{dx}(\cosh x) = \sinh x.$$

$$\frac{d}{dx}(\tanh x) = \text{sech}^2 x.$$

$$\frac{d}{dx}(\coth x) = -\text{cosech}^2 x.$$

$$\frac{d}{dx}(\text{sech } x) = -\text{sech } x \tanh x.$$

$$\frac{d}{dx}(\text{cosech } x) = -\text{cosech } x \coth x.$$

$$\frac{d}{dx}(\sinh^{-1}x) = \frac{1}{\sqrt{x^2+1}}.$$

$$\frac{d}{dx}(\cosh^{-1}x) = \frac{1}{\sqrt{x^2-1}}.$$

$$\frac{d}{dx}(\tanh^{-1}x) = \frac{1}{1-x^2}.$$

$$\frac{d}{dx}(\coth^{-1}x) = \frac{1}{1-x^2}$$

$$\frac{d}{dx}(\text{sech}^{-1}x) = -\frac{1}{x\sqrt{1-x^2}}.$$

$$\frac{d}{dx}(\text{cosech}^{-1}x) = -\frac{1}{x\sqrt{1+x^2}}.$$

$$\frac{d}{dx}(\sinh^{-1}u) = \frac{1}{\sqrt{u^2+1}} \frac{du}{dx}.$$

$$\frac{du}{dx}(\cosh^{-1}u) = \frac{1}{\sqrt{u^2-1}} \cdot \frac{du}{dx}.$$

$$\frac{d}{dx}(\tanh^{-1}u) = \frac{1}{1-u^2} \cdot \frac{du}{dx}.$$

$$\frac{d}{dx}(\text{sech}^{-1}u) = -\frac{1}{u\sqrt{1-u^2}} \cdot \frac{du}{dx}.$$

$$\frac{d}{dx}\int_a^x f(x)\,dx = f(x).$$

$$\frac{d}{dx}(u_1, u_2 \ldots, u_n) = (u_1, u_2 \ldots, u_n)$$

$$\left[\frac{du_1/dx}{u_1} + \frac{du_2/dx}{u_2} + \frac{du_n/dx}{u_n}\right]$$

6

Integral Calculus

Fundamentals :

$\int d f(x) = f(x) + c.$

$d \int f(x) dx = f(x)\, dx.$

$\int a f(x) dx = a \int f(x)\, dx.$

$\int o\, dx = c.$

$\int x^n dx = \dfrac{x^{n+1}}{n+1} + c \qquad n \neq -1.$

$\int \dfrac{1}{x}\, dx = \log_e x + c.$

$\int e^x dx = e^x + c.$

$\int a^x dx = \dfrac{a^x}{\log_e a} + c,\ a > o.$

$a \neq 1.$

$\int f(y)\, dx = \int \dfrac{f(y)\, dy.}{dy/dx}$

$\int u\, dv = uv - \int v\, du$

$\int (u \pm v) dx = \int u dx \pm \int v dx.$

$\int u \dfrac{dv}{dx} dx = uv - \int v \dfrac{du}{dx} dx.$

$\int \sin x\, dx = -\cos x + c.$

$\int \cos x\, dx = \sin x + c.$

$\int \tan x\, dx = \log_e \sec x + c.$

$= -\log_e \cos x + c.$

$\int \cot x\, dx = \log_e \sin x + c.$

$= -\log_e \operatorname{cosec} x.$

$\int \sec x\, dx = \log_e \tan \left(\dfrac{x}{2} + \dfrac{\pi}{4}\right) + c$

$= \log_e (\sec x + \tan x) + c.$

$\int \operatorname{cosec} x\, dx = \log_e \tan \dfrac{x}{2} + c$

$= \log_e (\operatorname{cosec} x - \cot x) + c.$

$\int \sinh x\, dx = \cosh x + c.$

$\int \cosh x\, dx = \sinh x + c.$

$\int \operatorname{sech} x \tanh x\, dx = -\operatorname{sech} x + c$

$\int \operatorname{cosech} x \coth x\, dx = -\operatorname{coseh} x + c$

$\int \tanh x\, dx = \log_e \cosh x + c$

$= -\log_e \operatorname{sech} + c.$

$\int \coth x\, dx = \log_e \sinh x + c$

$= -\log_e \operatorname{cosech} x + c.$

$\int \operatorname{sech} x\, dx = \sin^{-1} \tanh x + c$

$= 2 \tan^{-1} e^x + c.$

$\int \operatorname{cosech} x\, dx = \log_e \tanh \dfrac{x}{2} + c.$

$\int \sin^2 x\, dx = \dfrac{x}{2} - \frac{1}{2} \sin x \cos x + c.$

$\int \cos^2 x\, dx = \dfrac{x}{2} + \frac{1}{2} \sin x \cos x + c.$

$\int \sec^2 x\, dx = \tan x + c.$

$\int \tan^2 x\, dx = \tan x - x + c.$

$\int \operatorname{cosec}^2 x\, dx = -\cot x + c.$

$$\int \frac{dx}{a^2+x^2}=\frac{1}{a}\tan^{-1}\frac{x}{a}+c=-\frac{1}{a}\cot^{-1}\frac{x}{a}+c.$$

$$\int \frac{dx}{a^2-x^2}=\frac{1}{a}\tanh^{-1}\frac{x}{a}=\frac{1}{2a}\log_e\frac{a+x}{a-x}, \text{ when } x<a$$

$$=\frac{1}{a}\coth^{-1}\frac{x}{a}=\frac{1}{2a}\log_e\frac{x+a}{x-a}, \text{ when } x>a.$$

$$\int \frac{dx}{\sqrt{a^2-x^2}}=\sin^{-1}\frac{x}{a} \text{ or} -\cos^{-1}\frac{x}{a}+c.$$

$$\int \frac{dx}{\sqrt{x^2\pm a^2}}=\log_e\left(x+\sqrt{x^2\pm a^2}\right)+c.$$

$$\int \frac{dx}{\sqrt{2ax-x^2}}=\cos^{-1}\left(\frac{a-x}{a}\right)+c$$

$$\int \frac{dx}{x\sqrt{x^2-a^2}}=\frac{1}{a}\sec^{-1}\left(\frac{x}{a}\right)+c=\frac{1}{a}\cos^{-1}\left(\frac{a}{x}\right)+c.$$

$$\int \frac{dx}{x\sqrt{a^2\pm x^2}}=-\frac{1}{a}\log_e\left(\frac{a+\sqrt{a^2\pm x^2}}{x}\right)+c.$$

$$\int \sqrt{a^2-x^2}\,dx=\tfrac{1}{2}\left(x\sqrt{a^2-x^2}+a^2\sin^{-1}\frac{x}{a}\right)+c.$$

$$=\frac{a^2}{2}\sin^{-1}\frac{x}{a}+\frac{x}{2}\sqrt{a^2-x^2}+c.$$

$$\int \sqrt{x^2\pm a^2}\,dx=\tfrac{1}{2}\{x\sqrt{x^2\pm a^2}\pm a^2\log_e(x+\sqrt{x^2\pm a^2})\}+c.$$

$$\int \frac{dx}{\sqrt{b+ax^2}}=\frac{1}{\sqrt{-a}}\sin^{-1}x\sqrt{\frac{-a}{b}}\ ;\ a<0,\ b>0$$

$$=\frac{1}{\sqrt{a}}\sinh^{-1}x\sqrt{\frac{a}{b}}\ ;\ a>0,\ b>0.$$

$$=\frac{1}{\sqrt{a}}\cosh^{-1}x\sqrt{\frac{a}{-b}}\ ;a>0,\ b<0.$$

$$=\frac{1}{\sqrt{a}}\log_e\left(x\sqrt{a}+\sqrt{b+ax^2}\right);\ a>0$$

$$\int\sqrt{b+ax^2}\,dx=\frac{x}{2}\sqrt{b+ax^2}+\frac{b}{2}\int\frac{dx}{\sqrt{b+ax^2}}+c$$

$$\int x\sqrt{a^2-x^2}\,dx=-\tfrac{1}{3}(a^2-x^2)\sqrt{a^2-x^2}+c$$

$$\int x\sqrt{x^2-a^2}\,dx=\tfrac{1}{3}(x^2-a^2)\sqrt{x^2-a^2}+c$$

$$\int x\sqrt{x^2+a^2}\,dx=\tfrac{1}{3}(x^2+a^2)\sqrt{x^2+a^2}+c$$

$$\int \log_e x\,dx=x(\log_e x-1)+c$$

$$\int \sin^2 x\,dx=\tfrac{1}{2}(x-\tfrac{1}{2}\sin 2x)+c$$

$$\int \cos^2 x\,dx=\tfrac{1}{2}(x+\tfrac{1}{2}\sin 2x)+c$$

$$\int \tan^2 x\,dx=\tan x-x+c$$

$$\int \cot^2 x\,dx=-\cot x-x+c$$

$$\int \tan^n x\,dx=\frac{\tan^{n-1}x}{n-1}-\int \tan^{n-2}x\,dx$$

$$\int \cot^n x\,dx=-\frac{\cot^{n-1}x}{n-1}-\int \cot^{n-2}x\,dx$$

$$\int \sin^m x\cos^n x\,dx=\frac{\sin^{m+1}x\cos^{n-1}x}{m+n}+\frac{n-1}{m+n}\int \sin^m x\cos^{n-2}x\,dx$$

$$=-\frac{\sin^{m-1}x\cos^{n+1}x}{m+n}+\frac{m-1}{m+n}\int \sin^{m-2}\cos^n x\,dx$$

Table of Integrals :

Expressions containing $(ax+b)$:

$$\int(ax+b)^n\,dx=\frac{1}{a(n+1)}(ax+b)^{n+1}\ ;\ n\neq 1$$

$$\int x(ax+b)^n\,dx=\frac{1}{a^2(n+4)}(ax+b)^{n+2}-\frac{b}{a^2(n+1)}(ax+b)^{n+1}\ ;$$

$$\int\frac{dx}{ax+b}=\frac{1}{a}\log_e(ax+b)$$

$$\int\frac{dx}{(ax+b)^2}=-\frac{1}{a(ax+b)}\ ;$$

$$\int\frac{dx}{(ax+b)^3}=-\frac{1}{2a(ax+b)^2}\ ;$$

$$\int\frac{x}{ax+b}\,dx=\frac{x}{a}-\frac{b}{a^2}\log(ax+b)\ ;$$

$$\int \frac{x}{(ax+b)^2}dx = \frac{b}{a^2(ax+b)} + \frac{1}{a^2}\log(ax+b) \;;$$

$$\int \frac{x}{(ax+b)^3}dx = \frac{b}{2a^2(ax+b)^2} - \frac{1}{a^2(ax+b)} \;;$$

$$\int \frac{x^2 dx}{ax+b} = \frac{1}{a^3}\left[\tfrac{1}{2}(ax+b)^2 - 2b(ax+b) + b^2\log(ax+b)\right]$$

$$\int \frac{x^2 dx}{(ax+b)^2} = \frac{1}{a^3}\left[(ax+b) - 2b\log(ax+b) - \frac{b^2}{(ax+b)}\right]$$

$$\int \frac{x^2 dx}{(ax+b)^2} = \frac{1}{a^3}\left[\log(ax+b) + \frac{2b}{(ax+b)} - \frac{b^2}{2(ax+b)^2}\right]$$

$$\int \frac{dx}{x(ax+b)} = \frac{1}{b}\log\frac{x}{ax+b} \;:$$

$$\int \frac{dx}{x^2(ax+b)} = -\frac{1}{bx} + \frac{a}{b^2}\log\frac{ax+b}{x} \;;$$

$$\int \frac{dx}{x^3(ax+b)} = \frac{2ax-b}{2b^2x^2} + \frac{a^2}{b^3}\log\frac{x}{ax+b} \;;$$

$$\int \frac{dx}{x(ax+b)^2} = \frac{1}{b(ax+b)} - \frac{1}{b^2}\log\frac{ax+b}{x} \;;$$

$$\int \frac{dx}{x(ax+b)^3} = \frac{1}{b^3}\left[\frac{1}{2}\left(\frac{ax+2b}{ax+b}\right)^2 + \log\frac{x}{ax+b}\right];$$

$$\int \frac{dx}{x^2(ax+b)^2} = -\frac{b+2ax}{b^2x(ax+b)} + \frac{2a}{b^3}\log\frac{ax+b}{x} \;;$$

$$\int \sqrt{ax+b}\,dx = \frac{2}{3a}\sqrt{(ax+b)^3} \;;$$

$$\int x\sqrt{ax+b}\,dx = \frac{2(3ax-2b)}{15a^2}\sqrt{(ax+b)^3} \;;$$

$$\int x^n\sqrt{ax+b}\,dx = 2\frac{2}{a^{n+1}}\int u^2(u^2-b)^n\,du \;; \qquad u = \sqrt{ax+b}$$

$$\int \frac{\sqrt{ax+b}}{x}dx = 2\sqrt{ax+b} + b\int \frac{dx}{x\sqrt{ax+b}} \;;$$

$$\int \frac{dx}{\sqrt{ax+b}} = \frac{2\sqrt{ax+b}}{a} \;;$$

$$\int \frac{x\,dx}{\sqrt{ax+b}} = \frac{2(ax-2b)}{3a^2}\sqrt{ax+b} \;;$$

$$\int \frac{x^n dx}{\sqrt{ax+b}} = \frac{2}{a^{n+1}} \int (u^2-b)^n \, du : \qquad u=\sqrt{ax+b}.$$

$$\int \frac{dx}{x\sqrt{ax+b}} = \frac{1}{\sqrt{b}} \log \frac{\sqrt{ax+b}-\sqrt{b}}{\sqrt{ax+b}+\sqrt{b}} : \text{ for } b>o.$$

$$= -\frac{2}{\sqrt{b}} \tanh^{-1} \sqrt{\frac{ax+b}{b}} \; ; \text{ for } b>o .$$

$$= \frac{2}{\sqrt{-b}} \tan^{-1} \sqrt{\frac{ax+b}{-b}} \; ; \text{ for } b<o.$$

$$\int \frac{dx}{x^2\sqrt{ax+b}} = -\frac{\sqrt{ax+b}}{bx} - \frac{a}{2b} \int \frac{dx}{x\sqrt{ax+b}} \; ;$$

$$\int \frac{dx}{x^n (ax+b)^m} = -\frac{1}{b^{m+n-1}} \int \frac{(u-a)^{m+n=2}}{u^m} du \; ; \; u=\frac{ax+b}{x}.$$

$$\int (ax+b)^{\pm n/2} dx = \frac{2(ax+b)^{\frac{2\pm n}{2}}}{a(2\pm n)} \; ;$$

$$\int x(ax+b)^{\pm n/2} dx = \frac{2}{a^2}\left[\frac{(ax+b)^{\frac{4\pm n}{2}}}{4\pm n} - \frac{b(ax+b)^{\frac{2\pm n}{2}}}{2\pm n}\right]:$$

$$\int \frac{dx}{x(ax+b)^{n/2}} = \frac{1}{b}\int \frac{dx}{x(ax+b)^{n-2/2}} - \frac{a}{b}\int \frac{dx}{(ax+b)^{n/2}}$$

$$\int \frac{x^n dx}{\sqrt{ax+b}} = \frac{2x^n\sqrt{ax+b}}{(2n+1)a} - \frac{2nb}{(2n+1)a}\int \frac{x^{n-1}\, dx}{\sqrt{ax+b}} \; ;$$

$$\int \frac{dx}{x^n\sqrt{ax+b}} = \frac{-\sqrt{ax+b}}{(n-1)bx^{n-1}} - \frac{(2n-3)a}{(2n-2)b}\int \frac{dx}{x^{n-1}\sqrt{ax+b}} \; ;$$

$$\int \frac{(ax+b)^{n/2}}{x} dx = a\int (ax+b)^{n-2/2} dx + b\int \frac{(ax+b)^{n-2/2}}{x} dx \; ;$$

Expressions Containing $(x^2 \pm p^2)$, and $(p^2 \pm x^2)$:

$$\int \sqrt{x^2 \pm p^2}\; dx = \tfrac{1}{2}[x\sqrt{x^2 \pm p^2} \mp p^2 \log (x+\sqrt{x^2 \pm p^2})] \; ;$$

$$\int \frac{dx}{\sqrt{x^2 \pm p^2}} = \log (x+\sqrt{x^2 \pm p^2}) \; ;$$

$$\int \frac{dx}{x\sqrt{x^2-p^2}} = \frac{1}{p} \cos^{-1}\left(\frac{p}{x}\right) = -\frac{1}{p}\sin^{-1}\left(\frac{p}{x}\right) \; ;$$

$$\int \frac{dx}{p^2+x^2} = \frac{1}{p}\tan^{-1}\left(\frac{x}{p}\right) = -\frac{1}{p}\cot^{-1}\left(\frac{x}{p}\right)$$

$$\int \frac{dx}{p^2-x^2} = \frac{1}{2p}\log\frac{p+x}{p-x} = \frac{1}{p}\tanh^{-1}\left(\frac{x}{p}\right)$$

$$\int \frac{dx}{\sqrt{p^2-x^2}} = \sin^{-1}\left(\frac{x}{p}\right) = -\cos^{-1}\left(\frac{x}{p}\right)$$

$$\int \sqrt{p^2-x^2}\,dx = \tfrac{1}{2}\left[x\sqrt{p^2-x^2}+p^2\sin^{-1}\left(\frac{x}{p}\right)\right]$$

$$\int \frac{dx}{x\sqrt{p^2\pm x^2}} = -\frac{1}{p}\log\left(\frac{p+\sqrt{p^2\pm x^2}}{x}\right)$$

Expressions Containing (ax^2+c) and (ax^n+c) :

$$\int \frac{dx}{(ax^2+c)} = \frac{1}{\sqrt{ac}}\tan^{-1}\left(x\sqrt{\frac{a}{c}}\right) \text{ when } a>0 \text{ and } c>0$$

$$= \frac{1}{2\sqrt{-ac}}\log\frac{\sqrt{c}+x\sqrt{-a}}{\sqrt{c}-x\sqrt{-a}} \text{ when a}<0 \text{ and } c>0$$

$$= \frac{1}{2\sqrt{-ac}}\log\frac{x\sqrt{a}-\sqrt{-c}}{x\sqrt{a}+\sqrt{-c}} \text{ when } a>0 \text{ and } c<0$$

$$\int \frac{dx}{x(ax^2+c)} = \frac{1}{2c}\log\frac{x^2}{ax^2+c}$$

$$\int \frac{dx}{x^2(ax^2+c)} = -\frac{1}{cx} - \frac{a}{c}\int\frac{dx}{ax^2+c}$$

$$\int \frac{x\,dx}{(ax^2+c)} = \frac{1}{2a}\log(ax^2+c)$$

$$\int \frac{x^2\,dx}{(ax^2+c)} = \frac{x}{a} - \frac{c}{a}\int\frac{dx}{ax^2+c}$$

$$\int \frac{dx}{(ax^2+c)^n} = \frac{1}{2(n-1)c}\cdot\frac{x}{(ax^2+c)^{n-1}} + \frac{2n-3}{2(n-1)c}\int\frac{dx}{(ax^2+c)^{n-1}}$$

when $n>1$

$$\int x(ax^2+c)^n\,dx = \frac{1}{2a}\,\frac{(ax^2+c)^{n+1}}{n+1} \quad \text{when } n\neq -1$$

$$\int \frac{x^n dx}{ax^2+c} = \frac{x^{n-1}}{a(n-1)} - \frac{c}{a}\int\frac{x^{n-2}}{ax^2+c}dx \quad \text{when } n\neq 1$$

$$\int \frac{x^2\,dx}{(ax^2+c)^n} = -\frac{1}{2(n-1)a}\cdot\frac{x}{(ax^2+c)^{n-1}} + \frac{1}{2(n-1)a}\int\frac{dx}{(ax^2+c)^{n-1}}$$

$$\int \frac{dx}{x^2(ax^2+c)^n} = \frac{1}{c}\int \frac{dx}{x^2(ax^2+c)^{n-1}} - \frac{a}{c}\int \frac{dx}{(ax^2+c)^n}$$

$$\int \sqrt{ax^2+c}\;dx = \frac{x}{2}\sqrt{ax^2+c} + \frac{c}{2\sqrt{a}}\log\left(x\sqrt{a}+\sqrt{ax^2+c}\right)$$

when $a>0$

$$= \frac{x}{2}\sqrt{ax^2+c} + \frac{c}{2\sqrt{-a}}\sin^{-1}\left(x\sqrt{\frac{-a}{c}}\right)$$

when $a<0$

$$\int \frac{dx}{\sqrt{ax^2+c}} = \frac{1}{\sqrt{a}}\log\left(x\sqrt{a}+\sqrt{ax^2+c}\right) \quad \text{when } a>0$$

$$= \frac{1}{\sqrt{-a}}\sin^{-1}\left(x\sqrt{\frac{-a}{c}}\right) \quad \text{when } a<0$$

$$\int x\sqrt{ax^2+c}\;dx = \frac{1}{3a}(ax^2+c)^{3/2}$$

$$\int \frac{x\,dx}{\sqrt{ax^2+c}} = \frac{1}{a}\sqrt{ax^2+c}$$

$$\int \frac{x^2dx}{\sqrt{ax^2+c}} = \frac{x}{a}\sqrt{ax^2+c} - \frac{1}{a}\int\sqrt{ax^2+c}\;dx$$

$$\int \frac{\sqrt{ax^2+c}}{x}dx = \sqrt{ax^2+c} + \sqrt{c}\log\frac{\sqrt{ax^2+c}-\sqrt{c}}{x} \quad \text{when } c>0$$

$$= \sqrt{ax^2+c} - \sqrt{-c}\tan^{-1}\frac{\sqrt{ax^2+c}}{\sqrt{-c}} \quad \text{when } c<0$$

$$\int \frac{dx}{x\sqrt{ax^2+c}} = \frac{1}{\sqrt{c}}\log\frac{\sqrt{ax^2+c}-\sqrt{c}}{x} \quad \text{when c} >0$$

$$= \frac{1}{\sqrt{-c}}\sec^{-1}\left(x\sqrt{-\frac{a}{c}}\right) \quad \text{when } c<0$$

$$\int \frac{dx}{x^2\sqrt{ax^2+c}} = -\frac{\sqrt{ax^2+c}}{cx}$$

$$\int \frac{dx}{(ax^2+c)^{3/2}} = \frac{x}{c\sqrt{ax^2+c}}$$

$$\int x(ax^2+c)^{3/2}\,dx = \frac{1}{5a}(ax^2+c)^{5/2}$$

$$\int x^2(ax^2+c)^{3/2}\,dx = \frac{x^3}{6}(ax^2+c)^{3/2} + \frac{c}{2}\int x^2\sqrt{ax^2+c}\;dx$$

$$\frac{x\,dx}{(ax^2+c)^{3/2}}=-\frac{1}{a\sqrt{ax^2+c}}$$

$$\int\frac{dx}{x(ax^n+c)}=\frac{1}{cn}\log\frac{x^n}{ax^n+c}$$

$$\int\frac{dx}{(ax^n+c)^m}=\frac{1}{c}\int\frac{dx}{(ax^n+c)^{m-1}}-\frac{a}{c}\int\frac{x^n dx}{(ax^n+c)^m}$$

$$\int\frac{dx}{x\sqrt{ax^n+c}}=\frac{1}{n\sqrt{c}}\log\frac{\sqrt{ax^n+c}-\sqrt{c}}{\sqrt{ax^n+c}+\sqrt{c}}\quad\text{when } c>0$$

$$\int\frac{dx}{x\sqrt{ax^n+c}}=\frac{2}{n\sqrt{-c}}\sec^{-1}\sqrt{\frac{-ax^n}{c}}\quad\text{when } c<0$$

Expressions Containing (ax^2+bx+c) :

$$\int\frac{dx}{(ax^2+bx+c)}=\frac{1}{\sqrt{b^2-4ac}}\log\frac{2ax+b-\sqrt{b^2-4ac}}{2ax+b+\sqrt{b^2-4ac}}\quad\text{when } b^2>4ac$$

$$=\frac{2}{\sqrt{4ac-b^2}}\tan^{-1}\frac{2ax+b}{\sqrt{4ac-b^2}}\quad\text{when } b^2<4ac$$

$$=-\frac{2}{2ax+b}\quad\text{when } b^2=4ac$$

$$\int\frac{x\,dx}{(ax^2+bx+c)}=\frac{1}{2a}\log(ax^2+bx+c)-\frac{b}{2a}\int\frac{dx}{(ax^2+bx+c)}$$

$$\int\frac{x^2dx}{(ax^2+bx+c)}=\frac{x}{a}-\frac{b}{2a^2}\log(ax^2+bx+c)+\frac{b^2-2ac}{2a^2}\cdot\int\frac{dx}{(ax^2+bx+c)}$$

$$\int\frac{dx}{x(ax^2+bx+c)}=\frac{1}{2c}\log\left(\frac{x^2}{ax^2+bx+c}\right)-\frac{b}{2c}\int\frac{dx}{(ax^2+bx+c)}$$

$$\int\frac{dx}{x^2(ax^2+bx+c)}=\frac{b}{2c^2}\log\left(\frac{ax^2+bx+c}{x^2}\right)-\frac{1}{cx}+\left(\frac{b^2}{2c^2}-\frac{a}{c}\right)\int\frac{dx}{(ax^2+bx+c)}$$

$$\int\frac{dx}{\sqrt{ax^2+bx+c}}=\frac{1}{\sqrt{a}}\log\left(2ax+b+2\sqrt{a}\sqrt{ax^2+bx+c}\right)\quad\text{when } a>0$$

$$=\frac{1}{\sqrt{-a}}\sin^{-1}\frac{-(2ax+b)}{\sqrt{b^2-4ac}}\quad\text{when } a<0$$

$$\int \frac{xdx}{\sqrt{ax^2+bx+c}} = \frac{\sqrt{ax^2+bx+c}}{a} - \frac{b}{2a}\int \frac{dx}{\sqrt{ax^2+bx+c}}$$

$$\int \sqrt{ax^2+bx+c}\, dx = \frac{2ax+b}{4a}\sqrt{ax^2+bx+c} + \frac{4ac-b^2}{8a}\int \frac{dx}{\sqrt{ax^2+bx+c}}$$

$$\int x\sqrt{ax^2+bx+c}\, dx = \frac{(ax^2+bx+c)^{3/2}}{3a} - \frac{b}{2a}\int \sqrt{ax^2+bx+c}\, dx$$

$$\int x^2\sqrt{ax^2+bx+c}\, dx = \left(x - \frac{5b}{6a}\right)\frac{(ax^2+bx+c)^{3/2}}{4a} + \frac{(5b^2-4ac)}{16a^2}\int \sqrt{ax^2+bx+c}\, dx$$

$$\int \frac{dx}{x\sqrt{ax^2+bx+c}} = -\frac{1}{\sqrt{c}} \log\left(\frac{\sqrt{ax^2+bx+c}+\sqrt{c}}{x} + \frac{b}{2\sqrt{c}}\right) \quad \text{where } c>0$$

$$= \frac{1}{\sqrt{-c}} \sin^{-1} \frac{bx+2c}{x\sqrt{b^2-4ac}} \quad \text{where } c<0$$

$$\int \frac{dx}{(ax^2+bx+c)^{3/2}} = -\frac{2(2ax+b)}{(b^2-4ac)\sqrt{ax^2+bx+c}} \quad \text{when } b^2 \neq 4ac$$

$$= -\frac{1}{2a^{3/2}\left(x+\frac{b}{2a}\right)^2} \quad \text{when } b^2=4ac$$

Expressions Containing (sin ax) :

$$\int \sin ax\, dx = -\frac{1}{a}\cos ax$$

$$\int \sin^2 ax\, dx = \frac{x}{2} - \frac{\sin 2ax}{4a}$$

$$\int \sin^3 ax\, dx = -\frac{1}{a}\cos ax \frac{1}{3a}\cos^3 ax$$

$$\int \sin^n ax\, dx = -\frac{\sin^{n-1} ax \cos ax}{na} + \frac{n-1}{n}\int \sin^{n-2} ax\, dx$$

$$\int \frac{dx}{\sin ax} = \frac{1}{a}\log\tan\frac{ax}{2} = \frac{1}{a}\log(\operatorname{cosec} ax - \cot ax).$$

$$\int \frac{dx}{\sin^2 ax} = \int \operatorname{cosec}^2 ax\, dx = -\frac{1}{a}\cot ax$$

$$\int \frac{dx}{\sin^n ax} = -\frac{1}{a(n-1)}\cdot\frac{\cos ax}{\sin^{n-1} ax} + \frac{n-2}{n-1}\int \frac{dx}{\sin^{n-2} ax} \quad \text{when } n>1$$

$$\int \frac{dx}{1+\sin ax} = -\frac{1}{a} \tan\left(\frac{\pi}{4} - \frac{ax}{2}\right)$$

$$\int \frac{dx}{1-\sin ax} = \frac{1}{a} \tan\left(\frac{\pi}{4} + \frac{ax}{2}\right)$$

$$\int \sin ax \sin bx \, dx = \frac{\sin (a-b)x}{2(a-b)} - \frac{\sin (a+b)x}{2(a+b)} \quad \text{when } a^2 \neq b^2$$

$$\int \frac{dx}{b+c \sin ax} = \frac{-2}{a\sqrt{b^2-c^2}} \tan^{-1} \left\{ \sqrt{\frac{b-c}{b+c}} \tan\left(\frac{\pi}{4} - \frac{ax}{2}\right) \right\} \quad b^2 > c^2$$

$$= \frac{-1}{a\sqrt{c^2-b^2}} \log \frac{c+b \sin ax + \sqrt{c^2-b^2} \cos ax}{b+c \sin ax} \; ; \; b^2 < c^2$$

Expressions Containing (cos ax) :

$$\int \cos ax \, dx = \frac{1}{a} \sin ax$$

$$\int \cos^2 ax \, dx = \frac{x}{2} + \frac{\sin^2 ax}{4a}$$

$$\int \cos^3 ax \, dx = \frac{1}{a} \sin ax - \frac{1}{3a} \sin^3 ax$$

$$\int \cos^n ax \, dx = \frac{\cos^{n-1} ax \sin ax}{na} + \frac{n-1}{n} \int \cos^{n-2} ax \, dx$$

$$\int \frac{dx}{\cos ax} = \frac{1}{a} \log \tan\left(\frac{ax}{2} + \frac{\pi}{4}\right) = \frac{1}{a} \log (\sec ax + \tan ax)$$

$$\int \frac{dx}{\cos^2 ax} = \frac{1}{a} \tan ax$$

$$\int \frac{dx}{\cos^n ax} = \frac{1}{a(n-1)} \cdot \frac{\sin ax}{\cos^{n-1} ax} + \frac{n-2}{n-1} \int \frac{dx}{\cos^{n-2} ax} \quad \text{when } n > 1$$

$$\int \sin^m ax \cos^n ax \, dx = -\frac{\sin^{m-1} ax \cos^{n+1} ax}{a(m+n)}$$

$$+ \frac{m-1}{m+n} \int \sin^{m-2} ax \cos^n ax \, dx \; ; \quad \text{when } m, n > 0$$

$$= \frac{\sin^{m+1} ax \cos^{n-1} ax}{a(m+n)} + \frac{n-1}{m+n} \int \sin^m ax \cos^{n-2} ax \, dx$$

when $m, n > 0$

$$\int \frac{dx}{\sin^m ax \cos^n ax} = \frac{1}{a(n-1)} \cdot \frac{1}{\sin^{m-1} ax \cos^{n-1} ax} + \frac{m+n-2}{n-1} \int \frac{dx}{\sin^m ax \cos^{n-2} ax}$$

$$= -\frac{1}{a(m-1)} \cdot \frac{1}{\sin^{m-1} ax \cos^{n-1} ax} + \frac{m+n-2}{m-1} \int \frac{dx}{\sin^{m-2} ax \cos^n ax}$$

$$\int \frac{\sin^m ax}{\cos^n ax} dx = \frac{\sin^{m+1} ax}{a(n-1) \cos^{n-1} ax} - \frac{m-n+2}{n-1} \int \frac{\sin^m ax}{\cos^{n-2} ax} dx;$$

when $n \neq 1$ and $m, n > 0$

$$\int \frac{\cos^n ax}{\sin^m ax} dx = \frac{-\cos^{n+1} ax}{a(m-1) \sin^{m-1} ax} + \frac{m-n-2}{m-1} \int \frac{\cos^n ax}{\sin^{m-2} ax} dx;$$

when $m \neq 1$ and $m, n > 0$

$$\int \frac{\cos ax \, dx}{b + c \sin ax} = \frac{1}{ac} \log (b + c \sin ax)$$

$$\int \frac{\sin ax \, dx}{b + c \cos ax} = -\frac{1}{ac} \log (b + c \cos ax)$$

$$\int \frac{dx}{b \sin ax + c \cos ax} = \frac{1}{a\sqrt{b^2+c^2}} \left\{ \log \tan \tfrac{1}{2} \left(ax + \tan^{-1} \frac{c}{b} \right) \right\}$$

$$\int \frac{dx}{b^2 \cos^2 ax + c^2 \sin^2 ax} = \frac{1}{abc} \tan^{-1} \left(\frac{c \tan ax}{b} \right)$$

$$\int \frac{dx}{b^2 \cos^2 ax - c^2 \sin^2 ax} = \frac{1}{2abc} \log \frac{b \cos ax + c \sin ax}{b \cos ax - c \sin ax}$$

$$\int \frac{\sin ax \cos ax \, dx}{b \cos^2 ax + c \sin^2 ax} = \frac{1}{2a(c-b)} \log (b \cos^2 ax + c \sin^2 ax)$$

$$\int \frac{\sin^2 ax \, dx}{b + c \cos^2 ax} = \frac{1}{ac} \sqrt{\frac{b+c}{b}} \tan^{-1} \left\{ \sqrt{\frac{b}{b+c}} \tan ax \right\} - \frac{x}{c}$$

$$\int \frac{dx}{1 + \cos ax} = \frac{1}{a} \tan \frac{ax}{2}$$

$$\int \frac{dx}{1 - \cos ax} = -\frac{1}{a} \cot \frac{ax}{2}$$

$$\int \cos ax \cos bx \, dx = \frac{\sin (a-b)x}{2(a-b)} + \frac{\sin (a+b)x}{2(a+b)}$$

when $a^2 \neq b^2$

$$\int \frac{dx}{b+c\cos ax}=\frac{1}{a\sqrt{b^2-c^2}}\tan^{-1}\left(\frac{\sqrt{b^2-c^2}\sin ax}{c+b\cos ax}\right)\quad\text{when } b^2>c^2$$

$$=\frac{1}{a\sqrt{c^2-b^2}}\tanh^{-1}\left\{\frac{\sqrt{c^2-b^2}\sin ax}{c+b\cos ax}\right\}\quad\text{when } b^2<c^2$$

Expressions Containing sin ax and cos ax :

$$\int \sin ax\cos bx\,dx=-\frac{1}{2}\left\{\frac{\cos(a-b)x}{a-b}+\frac{\cos(a+b)x}{a+b}\right\}\quad\text{when } a^2\neq b^2$$

$$\int \sin^n ax\cos ax\,dx=\frac{1}{a(n+1)}\sin^{n+1}ax\quad\text{when } n\neq-1$$

$$\int \cos^n ax\sin ax\,dx=-\frac{1}{a(n+1)}\cos^{n+1}ax\quad\text{when } n\neq-1$$

$$\int \frac{\sin ax}{\cos ax}\,dx=-\frac{1}{a}\log\cos ax$$

$$\int \frac{\cos ax}{\sin ax}\,dx=\frac{1}{a}\log\sin ax$$

$$\int \sin^2 ax\cos^2 ax\,dx=\frac{x}{8}-\frac{\sin 4ax}{32a}$$

$$\int \frac{dx}{\sin ax\cos ax}=\frac{1}{a}\log\tan ax$$

$$\int \frac{dx}{\sin^2 ax\cos^2 ax}=\frac{1}{a}(\tan ax-\cot ax)$$

$$\int \frac{\sin^2 ax}{\cos ax}\,dx=\frac{1}{a}\left\{-\sin ax+\log\tan\left(\frac{ax}{2}+\frac{\pi}{4}\right)\right\}$$

$$\int \frac{\cos^2 ax}{\sin ax}\,dx=\frac{1}{a}\left\{\cos ax+\log\tan\frac{ax}{2}\right\}$$

Expressions Containing tan ax:

$$\int \tan ax\,dx=-\frac{1}{a}\log\cos ax$$

$$\int \tan^2 ax\,dx=\frac{1}{a}\tan ax-x$$

$$\int \tan^3 ax\, dx = \frac{1}{2a} \tan^2 ax + \frac{1}{a} \log \cos ax$$

$$\int \tan^n ax\, dx = \frac{1}{a\,(n-1)} \tan^{n-1} ax - \int \tan^{n-2} ax\, dx \text{ ; when } n>1$$

$$\int \cot x\, dx = \log \sin x = -\log \operatorname{cosec} x$$

$$\int \cot^2 ax\, dx = \int \frac{dx}{\tan^2 ax} = -\frac{1}{a} \cot ax - x$$

$$\int \cot^3 ax\, dx = -\frac{1}{2a} \cot^2 ax - \frac{1}{a} \log \sin ax$$

$$\int \cot^n ax\, dx = \int \frac{dx}{\tan^n ax} = -\frac{1}{a(n-1)} \cot^{n-1} ax$$

$$-\int \cot^{n-2} ax\, dx \text{ : when } n>1$$

$$\int \frac{dx}{b+c \tan ax} = \int \frac{\cot ax\, dx}{b \cot ax + c} = \frac{1}{b^2+c^2} \cdot$$

$$\left\{ bx + \frac{c}{a} \log (b \cos ax + c \sin ax) \right\}$$

$$\int \frac{dx}{b+c \cot ax} = \int \frac{\tan ax\, dx}{b \tan ax + c} = \frac{1}{b^2+c^2} \cdot$$

$$\left\{ bx - \frac{c}{a} \log (c \cos ax + b \sin ax) \right\}$$

Expressions Containing sec ax and cosec ax :

$$\int \sec ax\, dx = \frac{1}{a} \log \tan\left(\frac{ax}{2} + \frac{\pi}{4} \right);$$

$$\int \sec^2 ax\, dx = \frac{1}{a} \tan ax \text{ ;}$$

$$\int \sec^n ax\, dx = \frac{1}{a(n-1)} \cdot \frac{\sin ax}{\cos^{n-1} ax} + \frac{n-2}{n-1} \int \sec^{n-2} ax\, dx;$$

when $n>1$

$$\int \operatorname{cosec} ax\, dx = \frac{1}{a} \log \tan \frac{ax}{2} \text{ ;}$$

$$\int \operatorname{cosec}^n ax\, dx = -\frac{1}{a} \cot ax;$$

$$\int \operatorname{cosec}^n ax\, dx = -\frac{1}{a(n-1)} \cdot \frac{\cos ax}{\sin^{n-1} ax} + \frac{n-2}{n-1} \int \operatorname{cosec}^{n-2} ax\, dx \text{ ;}$$

when $n>1$

Expressions Containing tan ax and sec ax or cot ax and cosec ax :

$$\int \tan ax \sec ax\, dx = \frac{1}{a} \sec ax\,;$$

$$\int \tan ax \sec^n ax\, dx = \frac{1}{an} \sec^n ax\,; \quad \text{when } n \neq 0$$

$$\int \tan^n ax \sec^2 ax\, dx = \frac{1}{a(n+1)} \tan^{n+1} ax; \text{ when } n \neq -1$$

$$\int \cot ax \operatorname{cosec} ax\, dx = -\frac{1}{a} \operatorname{cosec} ax\,;$$

$$\int \cot ax \operatorname{cosec}^n ax\, dx = -\frac{1}{an} \operatorname{cosec}^n ax\,; \text{ when } n \neq 0$$

$$\int \cot ax \operatorname{cosec}^2 ax\, dx = -\frac{1}{a(n+1)} \cot^{n+1} ax\,; \text{ when } n \neq -1$$

Expressions involving exponential and logarithmic functions

$$\int e^x\, dx = e^x\,; \quad \int a^x\, dx = \frac{a^x}{\log a}\,; \int e^{ax}\, dx = \frac{1}{a} e^{ax}\,;$$

$$\int b^{ax}\, dx = \frac{b^{ax}}{a \log b}\,; \int x e^{ax}\, dx = \frac{e^{ax}}{a^2} (ax-1)\,;$$

$$\int x b^{ax}\, dx = \frac{x b^{ax}}{a \log b} - \frac{b^{ax}}{a^2 (\log b)^2}\,;$$

$$\int x^2 e^{ax}\, dx = \frac{e^{ax}}{a^3} (a^2x^2 - 2ax + 2)\,;$$

$$\int x^n e^{ax}\, dx = \frac{e^{ax}}{a^{n+1}} \left\{ (ax)^n - n(ax)^{n-1} + n(n-1) (ax)^{n-2} - \ldots\ldots + (-1)^n n! \right\}; \quad \text{where } n \text{ is } a + \text{ve integer.}$$

$$\int x^n e^{-ax}\, dx = -\frac{e^{-ax}}{a^{n+1}} \{ (ax)^n + n(ax)^{n-1} + n(n-1)(ax)^{n-2} + \ldots\ldots + n! \}\,;$$

where n is $a+$ve integer.

$$\int \frac{e^{ax}}{x}\, dx = \log x + ax + \frac{(ax)^2}{2 \cdot 2!} + \frac{(ax)^3}{3 \cdot 3!} + \ldots\ldots$$

$$\int \frac{e^{ax}}{x^n}\,dx = \frac{1}{n-1}\left\{-\frac{e^{ax}}{x^{n-1}} + a\int \frac{e^{ax}}{x^{n-1}}\,dx\right\} \quad \text{when } n>1$$

$$\int \frac{dx}{be^{ax}+ce^{-ax}} = \frac{1}{a\sqrt{bc}}\tan^{-1}\left(e^{ax}\sqrt{\frac{b}{c}}\right)$$

$$\int e^{ax}\sin bx\,dx = \frac{e^{ax}}{a^2+b^2}(a\sin bx - b\cos bx)$$

$$\int e^{ax}\cos bx\,dx = \frac{e^{ax}}{a^3+b^2}(a\cos bx + b\sin bx)$$

$$\int \log ax\,dx = x\log ax - x$$

$$\int x\log ax\,dx = \frac{x^2}{2}\log ax - \frac{x^2}{4}$$

$$\int x^2\log ax\,dx = \frac{x^3}{3}\log ax - \frac{x^3}{9}$$

$$\int (\log ax)^2 dx = x(\log ax)^2 - 2x\log ax + 2x$$

$$\int (\log ax)^n\,dx = x(\log ax)^n - n\int (\log ax)^{n-1}\,dx$$

$$\int x^n\log ax\,dx = x^{n+1}\left\{\frac{\log ax}{n+1} - \frac{1}{(n+1)^2}\right\}. \text{ when } n \neq -1$$

$$\int x^n(\log ax)^m\,dx = \frac{x^{n+1}}{n+1}(\log ax)^m - \frac{m}{n+1}\int x^n(\log ax)^{m-1}\,dx$$

$$\int \frac{(\log ax)^n}{x}\,dx = \frac{(\log ax)^{n+1}}{n+1}\ ;\ \text{when } n \neq -1$$

$$\int \frac{dx}{x\log ax} = \log(\log ax);$$

$$\int \frac{dx}{x(\log ax)^n} = -\frac{1}{(n-1)(\log ax)^{n-1}};$$

$$\int \frac{dx}{\log ax} = \frac{1}{a}\left\{\log(\log ax) + \log ax + \frac{(\log ax)^2}{2\cdot 2\,!} + \frac{(\log ax)^3}{3\cdot 3\,!} + \ldots\right\}$$

$$\int \frac{x^n\,dx}{\log ax}$$

$$=\frac{1}{a^{n+1}}\left\{\log(\log ax)+(n+1)\log ax+\frac{(n+1)^2(\log ax)^2}{2\cdot 2!}+\dots\dots\right\}$$

$$\int\frac{x^n\,dx}{(\log ax)^m}=\frac{-x^{n+1}}{(m-1)(\log ax)^{m-1}}+\frac{n+1}{m-1}\int\frac{x^n\,dx}{(\log ax)^{m-1}}$$

when $m\neq 1$

$$\int\sin(\log ax)\,dx=\frac{x}{2}\left\{\sin(\log ax)-\cos(\log ax)\right\};$$

$$\int\cos(\log ax)\,dx=\frac{x}{2}\left\{\sin(\log ax)+\cos(\log ax)\right\};$$

$$\int e^{ax}\log bx\,dx=\frac{1}{a}e^{ax}\log bx-\frac{1}{a}\int\frac{e^{ax}}{x}dx;$$

Expressions involving Algebraic and Trigonometric Functions

$$\int x\sin ax\,dx=\frac{1}{a^2}\sin ax-\frac{1}{a}x\cos ax$$

$$\int x\sin^2 ax\,dx=\frac{x^2}{4}-\frac{x\sin 2ax}{4a}-\frac{\cos 2ax}{8a^2}$$

$$\int x^2\sin ax\,dx=\frac{2x}{a^2}\sin ax+\frac{2}{a^3}\cos ax-\frac{x^2}{a}\cos ax$$

$$\int x^2\sin^2 ax\,dx=\frac{x^3}{6}-\left(\frac{x^2}{4a}-\frac{1}{8a^3}\right)\sin 2ax-\frac{x\cos 2ax}{4a^2}$$

$$\int x^n\sin ax\,dx=-\frac{1}{a}x^n\cos ax+\frac{n}{a}\int x^{n-1}\cos ax\,dx$$

$$\int\frac{\sin ax\,dx}{x}=ax-\frac{(ax)^3}{3\cdot 3!}+\frac{(ax)^5}{5\cdot 5!}-\dots\dots$$

$$\int x\cos ax\,dx=\frac{1}{a^2}\cos ax+\frac{1}{a}x\sin ax$$

$$\int x\cos^2 ax\,dx=\frac{x^2}{4}+\frac{x\sin 2ax}{4a}+\frac{\cos 2ax}{8a^2}$$

$$\int x^2\cos ax\,dx=\frac{2x}{a^2}\cos ax-\frac{2}{a^3}\sin ax+\frac{x^2}{a}\sin ax$$

$$\int x^2\cos^2 ax\,dx=\frac{x^3}{6}+\left(\frac{x^2}{4a}-\frac{1}{8a^3}\right)\sin 2ax+\frac{x\cos 2ax}{4a^2}$$

$$\int x^n\cos ax\,dx=\frac{1}{a}x^n\sin ax-\frac{n}{a}\int x^{n-1}\sin ax\,dx$$

$$\int\frac{\cos ax\,dx}{x}=\log ax-\frac{(ax)^2}{2\cdot 2!}+\frac{(ax)^4}{4\cdot 4!}-\dots\dots$$

Expressions involving Inverse Trigonometric Functions :

$$\int \sin^{-1} ax\, dx = x \sin^{-1} ax + \frac{1}{a}\sqrt{1-(ax)^2}$$

$$\int x \sin^{-1} ax\, dx = \frac{x^2}{2} \sin^{-1} ax - \frac{1}{4a^2} \sin^{-1} ax + \frac{x}{4a} \sqrt{1-(ax)^2}$$

$$\int \frac{\sin^{-1} ax\, dx}{x} = ax + \frac{1}{2.3.3}(ax)^3 + \frac{1}{2.4.5.5}(ax)^5 + \ldots..$$

$$\int (\sin^{-1} ax)^2\, dx = x(\sin^{-1} ax)^2 - 2x + \frac{2}{a}\sqrt{1-(ax)^2}\, \sin^{-1} ax$$

$$\int \frac{\sin^{-1} ax\, dx}{x^2} = -\frac{1}{x} \sin^{-1} ax - a \log \left| \frac{1+\sqrt{1-(ax)^2}}{ax} \right|$$

$$\int \cos^{-1} ax\, dx = x \cos^{-1} ax - \frac{1}{a}\sqrt{1-(ax)^2}$$

$$\int x \cos^{-1} ax\, dx = \frac{x^2}{2} \cos^{-1} ax - \frac{1}{4a^2}\cos^{-1} ax - \frac{x}{4a}\sqrt{1-(ax)^2}$$

$$\int \frac{\cos^{-1} ax\, dx}{x} = \frac{\pi}{2} \log |\, ax \,| - ax - \frac{1}{2.3.3}(ax)^3 - \frac{1.3}{2.4.5.5}(ax)^5 - \ldots$$

$$\int (\cos^{-1} ax)^2\, dx = x(\cos^{-1} ax)^2 - 2x - \frac{2}{a}\sqrt{1-(ax)^2}\, \cos^{-1} ax$$

$$\int \frac{\cos^{-1} ax\, dx}{x^2} = -\frac{1}{x} \cos^{-1} ax + a \log \left| \frac{1+\sqrt{1-(ax)^2}}{ax} \right|$$

$$\int \tan^{-1} ax\, dx = x \tan^{-1} ax = \frac{1}{2a} \log (1+a^2x^2)$$

$$\int \cot^{-1} ax\, dx = x \cot^{-1} ax + \frac{1}{2a} \log (1+a^2x^2)$$

$$\int \sec^{-1} ax\, dx = x \sec^{-1} ax - \frac{1}{a} \log \left(ax + \sqrt{(ax)^2-1} \right)$$

$$\int \operatorname{cosec}^{-1} ax\, dx = x \operatorname{cosec}^{-1} ax + \frac{1}{a} \log \left(ax + \sqrt{(ax)^2-1} \right)$$

$$\int \frac{\tan^{-1} ax\, dx}{x^2} = -\frac{1}{x} \log \tan^{-1} ax - \frac{a}{2} \log \left(\frac{1+a^2x^2}{a^2x^2} \right)$$

$$\int \frac{\cot^{-1} ax\, dx}{x^2} = -\frac{1}{x} \cot^{-1} ax + \frac{a}{2} \log \left(\frac{1+a^2x^2}{a^2x^2} \right)$$

Fundamentals of Definite Integral :

$f(x)$=Function of x. Let $f(x)$ be continuous between limits $x=a$ to $x=b$.

$$\int_a^b f(x)\,dx = \Big[F(x)\Big]_a^b = F(b) - F(a)\ ;$$

$$\int_a^b kf(x)\,dx = k\int_a^b f(x)dx = k\{F(b) - F(a)\}\ ; \text{ when } K = a \text{ constant}$$

$$\int_a^b f(x)dx = -\int_b^a f(x)dx\ ;$$

$$\int_a^b f(x)dx = \int_a^c f(x)dx + \int_c^b f(x)dx\ ;$$

$$\int_a^b f(x)dx = (b-a)\,f(x^1)\ : \quad \text{where } x^1 \text{ lies between } a \text{ and } b.$$

$$\int_a^{\infty} f(x)dx = \lim_{b\to\infty} \int_a^b f(x)dx\ ;$$

$$\int_{-\infty}^b f(x)dx = \lim_{a\to\infty} \int_{-a}^b f(x)dx\ ;$$

$$\int_{-\infty}^{+\infty} f(x)dx = \int_{-\infty}^c f(x)dx + \int_c^{\infty} f(x)dx\ ;$$

$$\int_a^b f(x)dx = \lim_{c\to 0} \int_a^{b-c} f(x)dx$$

Table of Definite Integrals :

$$\int_0^{\infty} \frac{\sin^2 x}{x^2}\, dx = \frac{\pi}{2} \;;$$

$$\int_0^{\pi} \sin^2 ax\, dx = \int_0^{\pi} \cos^2 ax\, dx = \frac{\pi}{2} \;;$$

$$\int_0^{\infty} \frac{\sin ax}{x}\, dx = \frac{\pi}{2} \;;\ \text{when } a > 0$$

$$\int_0^{\infty} \frac{\sin x \cos ax}{x}\, dx = \frac{\pi}{2} \;;\ \text{when } -1 < a < 1$$

$$= \frac{\pi}{4} \;;\ \text{when } a = 1 \text{ or } -1$$

$$= 0 \;;\ \text{when } a > 1 \text{ or } a < -1$$

$$\int_0^{\frac{\pi}{2}} \sin^n x\, dx = \int_0^{\frac{\pi}{2}} \cos^n x\, dx$$

$$= \frac{1\,.\,3\,.\,5 \ldots\ldots (n-1)}{2\,.\,4\,.\,6 \ldots\ldots (n)} \cdot \frac{\pi}{2} \qquad \text{when } n \text{ is even}$$

$$= \frac{2\,.\,4\,.\,6 \ldots\ldots (n-1)}{1\,.\,3\,.\,5 \ldots\ldots n} \qquad \text{when } n \text{ is odd}$$

$$\int_0^{\frac{\pi}{a}} \sin ax \cos ax\, dx = \int_0^{\pi} \sin ax \cos ax\, dx = 0$$

$$\int_0^{\pi} \sin ax \sin bx\, dx = \int_0^{\pi} \cos ax \cos bx\, dx = 0 \text{ when } b \neq a$$

$$\int_0^{\pi} \sin ax \cos bx \, dx = 0 \; ; \quad \text{when } (a-b) \text{ is even.}$$

$$= \frac{2a}{a^2 - b^2} \quad \text{when } (a-b) \text{ is odd}$$

$$\int_0^{\infty} \frac{\sin ax \sin bx}{x^2} dx = \frac{\pi}{2} a \qquad \text{when } b > a$$

$$\int_0^{\infty} \sin x^2 dx = \int_0^{\infty} \cos x^2 dx = \frac{1}{2}\sqrt{\frac{\pi}{2}} \; ;$$

$$\int_0^{\infty} x^n . e^{-ax} dx = \frac{n!}{a^{n+1}} \qquad \text{when } a > 0$$

$$\int_0^{\infty} x^{\frac{1}{2}} \; e^{-ax} dx = \frac{1}{2a}\sqrt{\frac{\pi}{a}}$$

$$\int_0^{\infty} e^{-ax} \sin \; bx \, dx = \frac{b}{a^2 + b^2} \qquad \text{when } a > 0$$

$$\int_0^{\infty} \frac{e^{-ax}}{x^{\frac{1}{2}}} dx = \sqrt{\frac{\pi}{a}}$$

$$\int_0^{\infty} e^{-ax} \cos bx \, dx = \frac{a}{a^2 + b^2} \qquad \text{when } a > 0$$

$$\int_0^{\infty} \frac{e^{-ax} \sin x}{x} dx = \cot^{-1} a \qquad \text{when } a > 0$$

$$\int_0^1 (\log x)^n dx = (-1)^n \cdot n\,!$$

$$\int_0^1 \frac{\log x}{1+x} dx = -\frac{\pi^2}{12}\ ; \qquad \int_0^1 \frac{\log x}{1-x} dx = -\frac{\pi^2}{6}\ ;$$

$$\int_0^1 \frac{\log x}{1-x^2} dx = -\frac{\pi^2}{8}\ ; \qquad \int_0^1 \log \frac{1+x}{1-x} \cdot \frac{dx}{x} = \frac{\pi^2}{4}\ ;$$

$$\int_0^1 \frac{\log x}{\sqrt{1-x^2}} = -\frac{\pi}{2} \log 2\ ; \qquad \int_0^1 \frac{dx}{\sqrt{\log\left(\frac{1}{x}\right)}} = \sqrt{\pi}\ ;$$

$$\int_0^1 \sqrt{\log \frac{1}{x}}\ dx = \frac{\sqrt{\pi}}{2}\ : \qquad \int_0^1 \left(\log \frac{1}{x}\right)^{-\frac{1}{2}} dx = \sqrt{\pi}\ ;$$

$$\int_0^1 \log(\log x)\ dx = \int_0^\infty e^{-x} \log x\ dx = -0{\cdot}57722\ ;$$

$$\int_0^1 \frac{x^a - x^b}{\log x} dx = \log \frac{1+a}{1+b}\ ; \qquad \int_0^1 \frac{1+x^2}{1+x^4} dx = \frac{\pi}{4}\sqrt{2}\ ;$$

$$\int_0^1 \frac{\log(1+x)}{x}\ dx = \frac{1}{1^2} - \frac{1}{2^2} + \frac{1}{3^2} - \frac{1}{4^2} + \ldots\ldots\ldots = \frac{\pi^2}{12}\ ;$$

$$\int_0^{\frac{\pi}{2}} \log \sin x\ dx = \int_0^{\frac{\pi}{2}} \log \cos x\ dx = -\frac{\pi}{2} \log_e 2\ ;$$

$$\int_0^{\pi} x \log \sin x \, dx = -\frac{\pi^2}{2} \log_e 2 \; ;$$

$$\int_0^{\pi/2} \frac{dx}{a+b\cos x} = \frac{\cos^{-1}\frac{b}{a}}{\sqrt{a^2-b^2}} \qquad \text{when } a>b$$

$$\int_0^{\pi} \frac{dx}{a+b\cos x} = \frac{\pi}{\sqrt{a^2-b^2}} \qquad \text{when } a>b>0$$

$$\int_0^{\pi} \frac{\log(1+\sin a \cos x)}{\cos x} \, dx = \pi a \; ;$$

$$\int_0^{\infty} \frac{\cos x \, dx}{x^{\frac{1}{2}}} = \int_0^{\infty} \frac{\sin x \, dx}{x^{\frac{1}{2}}} = \sqrt{\frac{\pi}{2}}$$

$$\int_0^{\infty} \frac{\cos ax \, dx}{1+x^2} = \frac{\pi}{2} e^{-a} \qquad \text{when } a>0$$

$$= \frac{\pi}{2} e^{a} \qquad \text{when } a<0$$

$$\int_0^{\infty} \frac{\cos ax - \cos bx}{x} \, dx = \log \frac{b}{a}$$

$$\int_0^{\pi} \frac{(a-b\cos x)\, dx}{a^2-2ab\cos x+b^2} = \frac{\pi}{2a} \qquad \text{when } a=b$$

$$= \frac{\pi}{a} \qquad \text{when } a^2>b^2$$

$$= 0 \qquad \text{when } a^2<b^2$$

$$\int_0^{\frac{\pi}{2}} \frac{dx}{a^2 \cos^2 x + b^2 \sin^2 x} = \frac{\pi}{2ab} ;$$

$$\int_0^{\frac{\pi}{2}} \frac{dx}{(a^2 \cos^2 x + b^2 \sin^2 x)^2} = \frac{\pi(a^2+b^2)}{4a^3 b^3}$$

$$\int_3^{\propto} \log \left(\frac{e^x+1}{e^x-1} \right) dx = \frac{\pi^2}{4} ;$$

$$\int_0^{\propto} \frac{e^{-ax} - e^{-bx}}{x} \, dx = \log \frac{b}{a} ;$$

$$\int_0^{\propto} e^{-x} \log x \, dx = -0{\cdot}57722$$

$$\int_0^{\propto} \frac{1}{x} \left(\frac{1}{1+x} - e^{-x} \right) dx = 0{\cdot}57722$$

$$\int_0^{\propto} \left(\frac{1}{1-e^{-x}} - \frac{1}{x} \right) dx = 0{\cdot}57722$$

$$\int_0^{\propto} x e^{-ax} \sin bx \, dx = \frac{2ab}{(a^2+b^2)^2} \qquad \text{when } a>0$$

$$\int_0^{\propto} x e^{-ax} \cos bx \, dx = \frac{a^2-b^2}{(a^2+b^2)} \quad \text{when } a>0)$$

$$\int_0^{\infty} e^{-ax} \sin h\, bx\, dx = \frac{b}{a^2-b^2} \qquad \text{when } a > |b| \geqq 0$$

$$\int_0^{\infty} e^{-ax} \cos h\, bx\, dx = \frac{a}{a^2-b^2} \qquad \text{when } a > |b| \geqq 0$$

$$\int_0^{\infty} x^2 e^{-ax} \sin bx\, dx = \frac{2b(3a^2-b^2)}{(a^2+b^2)^3} \qquad \text{when } a > 0$$

$$\int_0^{\infty} x^2 e^{-ax} \cos bx\, dx = \frac{2a(a^2-3b^2)}{(a^2+b^2)^3} \qquad \text{when } a > 0$$

$$\int_0^{\infty} x^n e^{-ax} \sin bx\, dx = \frac{i\,.\,n\,!\,\{a-ib)^{n+1}-(a+ib)^{n+1}\}}{2(a^2+b^2)^{n+1}} \qquad \text{when } a > 0$$

$$\int_0^{\infty} x^n e^{a-x} \cos bx\, dx = \frac{n\,!\,\{(a-ib)^{n+1}+(a+ib)^{n+1}\}}{2(a^2+b^2)^{n+1}} \qquad \text{when } a > 0$$

7

Numerical Analysis

Differences

Let y_0, y_1, y_2......y_n denote a set of values of any function

$$y=f(x) ;$$

Then, $\quad \triangle y_0=y_1-y_0,$

$$\triangle y_1=y_2-y_1 \ldots\ldots\ldots$$

$$\triangle y_{n-1}, =y_n-y_{n-1}, \qquad \triangle y_n=y_{n+1}=y_n$$

are called first differences. The second differences are :

$$\triangle^2 y_0=\triangle y_1-\triangle y_0,$$

$$\triangle^2 y_1=\triangle y_2-\triangle y_1 \ldots\ldots\ldots\ldots \text{etc. and so on.}$$

The following difference tables show how the difference of all orders are formed and arranged :

Diagonal Difference Table (for forward interpolation) :

x	y	$\triangle y$	$\triangle^2 y$	$\triangle^3 y$	$\triangle^4 y$	$\triangle^5 y$	$\triangle^6 y$	$\triangle^7 y$	$\triangle^8 y$
x_0	y_0								
		$\triangle y_0$							
x_1	y_1		$\triangle^2 y_0$						
		$\triangle y_1$		$\triangle^3 y_0$					
x_2	y_2		$\triangle^2 y_1$		$\triangle^4 y_0$				
		$\triangle y_2$		$\triangle^3 y_1$		$\triangle^5 y_0$			
x_3	y_3		$\triangle^2 y_2$		$\triangle^4 y_1$		$\triangle^6 y_0$		
		$\triangle y_3$		$\triangle^3 y_2$		$\triangle^5 y_1$		$\triangle^7 y_0$	
x_4	y_4		$\triangle^3 y_3$		$\triangle^4 y_2$		$\triangle^6 y_1$		$\triangle^8 y_0$
		$\triangle y_4$		$\triangle^3 y_3$		$\triangle^5 y_2$		$\triangle^7 y_1$	
x_5	y_5		$\triangle^4 y_4$		$\triangle_4 y_3$		$\triangle^6 y_2$		
		$\triangle y_5$		$\triangle^3 y_4$		$\triangle^5 y_3$			
x_6	y_6		$\triangle^5 y_5$		$\triangle^4 y_4$				
		$\triangle y_6$		$\triangle^3 y_5$					
x_7	y_7		$\triangle^6 y_6$						
		$\triangle y_7$							
x_8	y_8								

Newton's Formula (for forward interpolation) :

Let $\phi(x)$ be a polynomial of nth degree which may be written as

$$\phi(x)=a_0+a_1(x-x_0)+a_2(x-x_0)(x-x_1)+\ldots\ldots\ldots a_n(x-x_0)(x-x_1)\ldots\ldots(x+x_{n-1})$$

$$=y_0+\frac{\Delta y_0}{h}(x-x_0)+\frac{\Delta^2 y_0}{2!\,h^2}(x-x_0)(x-x_1)+\ldots\ldots\ldots$$

$$\ldots\ldots\ldots+\frac{\Delta^n y_0}{n!\,h^n}(x-x_0)(x-x_1)\ldots\ldots(x-x_{n-1}).$$

where $h=x_1-x_0=x_2-x_1\ldots\ldots\ldots=x_{n+1}-x_n$.

Substituting $\frac{x-x_0}{h}=u$,

$$\phi(x)=y_0+u\Delta y_0+\frac{u(u-1)}{2!}\Delta^2 y_0+\ldots\ldots\ldots\ldots\ldots\ldots$$

$$+\frac{u(u-1)(u-2)\ldots\ldots\ldots u-n-1}{n!}\Delta^n y_0$$

Horizontal Difference Table (for backward interpolation) :

x	y	$\Delta_1 y$	$\Delta_2 y$	$\Delta_3 y$	$\Delta_4 y$	$\Delta_5 y$	$\Delta_6 y$	$\Delta_7 y$	$\Delta_8 y$
x_0	y_0								
x_1	y_1	$\Delta_1 y_1$							
x_2	y_2	$\Delta_1 y_2$	$\Delta_2 y_2$						
x_3	y_3	$\Delta_1 y_3$	$\Delta_2 y_3$	$\Delta_3 y_3$					
x_4	y_4	$\Delta_1 y_4$	$\Delta_2 y_4$	$\Delta_3 y_4$	$\Delta_4 y_4$				
x_5	y_5	$\Delta_1 y_5$	$\Delta_2 y_5$	$\Delta_3 y_5$	$\Delta_4 y_5$	$\Delta_5 y_5$			
x_6	y_6	$\Delta_1 y_6$	$\Delta_2 y_6$	$\Delta_3 y_6$	$\Delta_4 y_6$	$\Delta_5 y_6$	$\Delta_6 y_6$		
x_7	y_7	$\Delta_1 y_7$	$\Delta_2 y_7$	$\Delta_3 y_7$	$\Delta_4 y_7$	$\Delta_5 y_7$	$\Delta_6 y_7$	$\Delta_7 y_6$	
x_8	y_8	$\Delta_1 y_8$	$\Delta_2 y_8$	$\Delta_3 y_8$	$\Delta_4 y_8$	$\Delta_5 y_8$	$\Delta_6 y_8$	$\Delta_7 y_8$	$\Delta_8 y$

Newton's Formula (for backward interpolation)

The polynomial $\phi(x)$ may be written in the form :

$$\phi(x)=a_0+a_1(x-x_n)+a_2(x-x_n)(x-x_{n-1})+\ldots\ldots\ldots$$

$$\ldots\ldots+a_n(x-x_n)(x-x_{n-1})\ldots\ldots(x-x_1)$$

$$=y_n+\frac{\triangle_1 y_n}{h}(x-x_n)+\frac{\triangle_2 y_n}{2!\,h^2}(x-x_n)(x-x_{n-1})+\ldots\ldots$$

$$\ldots\ldots\ldots+\frac{\triangle_n y_n}{n!\,h^n}(x-x_n)(x-x_{n-1})\ldots\ldots(x-x_1)$$

Putting $u=\frac{x-x_n}{h},\quad x=x_n+hu$

$\therefore\ \phi(x)=\phi(x_n+hu)=\psi(u)$, **say.**

$$\therefore\ \psi(u)=y_n+u\triangle_1 y_n\ \psi\ \frac{u(u+1)}{2!}\triangle_2 y_n+\ldots\ldots$$

$$+\frac{u(u+1)(u+2)\ldots\ldots(u+n-1)}{n!}\triangle_n y_n$$

Lagrange's Interpolation Formula :

Let $y=a_0(x-x_1)(x-x_2)\ldots\ldots(x-x_n)+a_1(x-x_0)(x-x_2)(x-x_3)$
$\ldots(x-x_n)+a_2(x-x_0)(x-x_1)(x-x_3)\ldots\ldots(x-x_n)+a_n(x-x_0)$
$(x-x_1)\ldots(x-x_{-1})$ be a polynomial.

Then,

$$y=\frac{(x-x_1)(x-x_2)(x-x_3)\ldots\ldots\ldots\ldots(x-x_n)}{(x_0-x_1)(x_0-x_2)(x_0-x_3)\ldots\ldots\ldots\ldots(x_0-x_n)}y_0$$

$$+\frac{(x-x_0)(x-x_2)(x-x_3)\ldots\ldots\ldots\ldots(x-x_n)}{(x_1-x_0)(x_1-x_2)(x_1-x_3)\ldots\ldots\ldots\ldots(x_1-x_n)}y_1$$

$$+\frac{(x-x_0)(x-x_1)(x-x_3)\ldots\ldots\ldots\ldots(x-x_n)}{(x_2-x_0)(x_2-x_1)(x_2-x_3)\ldots\ldots\ldots\ldots(x_2-x_n)}y_2+\ldots\ldots$$

$$\ldots\ldots\ldots+\frac{(x-x_0)(x-x_1)(x-x_2)\ldots\ldots\ldots\ldots(x-x_{n-1})}{(x_n-x_0)(x_n-x_1)(x_n-x_2)\ldots\ldots\ldots\ldots(x_n-x_{n-1})}y_n$$

The variables x and y are interchangeable and anyone of them may be taken as independent variable. The above formula is chiefly employed for finding out any value of a function when the given values of the independent variable are not equidistant.

Newton's General Interpolation Formula (for unequal intervals, with divided differences) :

$$y=y_0+(x-x_0)\ \delta(x_0, x_1)+(x-x_0)(x-x_1)\ \delta(x_0, x_1, x_2)+\ldots\ldots$$
$$\ldots\ldots+(x-x_0)(x-x_1)\ldots\ldots(x-x_{n-1})\ \delta(x_0, x_1, x_2\ldots\ldots x_n)$$
$$+(x-x_0)(x-x_1)\ldots\ldots(x-x_n)\ \delta(x_0, x_1, x_2,\ldots\ldots x_{n+1})$$

Gauss's Central Difference Formula (for forward interpolation) :

$$y=y_0+u\cdot\triangle y_0+u(u-1)\ \frac{\triangle^2 y_{-1}}{2!}$$

$$+u(u^2-1)\ \frac{\triangle^3 y_{-1}}{3!}+u(u^2-1)(u-2)\ \frac{\triangle^4 y_{-2}}{4!}$$

$$+u(u^2-1)(u^2-2^2)\frac{\triangle^5 y_{-2}}{5!}+\ldots\ldots\ldots$$

Gauss's Central Difference Formula (for backward interpolation):

$$y=y_0+u\Delta y_{-1}+u(u+1)\frac{\Delta y^2_{-1}}{2\,!}+u(u^2-1)\frac{\Delta^3 y_{-2}}{3\,!}$$

$$+u(u^2-1)(u^2-2^2)\frac{\Delta^4 y_{-2}}{4\,!}$$

$$+u(u^2-1)(u^2-2^2)\frac{\Delta^5 y_{-3}}{5\,!}+\ldots\ldots\ldots\ldots$$

Stirling's Interpolation Formula :

$$y=y_0+u\cdot\frac{\Delta y_{-1}+\Delta y_n}{2}+\frac{u^2}{2}\Delta^2 y_{-1}$$

$$+\frac{u(u^2-1)}{3\,!}\cdot\frac{\Delta^3 y_{-2}+\Delta^3 y_{-}}{2},$$

$$+\frac{u^2(u^2-1)}{4\,!}\Delta^4 y_{-2}$$

$$+\frac{u(u^2-1)(u^2-2^2)}{5\,!}\cdot\frac{\Delta^5 y_{-3}+\Delta^5 y_{-2}}{2}+\ldots\ldots\ldots\ldots$$

Legendre's Differential Equation :

$$(1-x^2)\frac{d^2y}{dx^2}-2x\frac{dy}{dx}+n(n+1)y=0,$$

the solution of which in the form of general equation is

$$y=a_0\left[1-\frac{n(n+1)}{2\,!}x^2+\frac{(n-2)n(n+1)(n+3)}{4\,!}x^4-\ldots\ldots\right]$$

$$+a_1\left[x-\frac{(n+2)(n-1)}{3\,!}x^3+\frac{(n+2)(n+4)(n-1)(n-3)}{5\,!}x^5-\ldots\ldots\right]$$

Hence, the polynominal

$$y=a_n\left[x^n-\frac{n(n-1)}{2(2n-1)}x^{n-2}+\frac{n(n-1)(n-2)(n-3)}{2\cdot 4(2n-1)(2n-3)}x^{n-4}\ldots\ldots\right]$$

This is a particular solution of Legendre's equation where a_n is arbitrary.

$$a_n=\frac{(2n-1)(2n-3)\ldots\ldots 3\cdot 1}{n\,!}=\frac{(2n)\,!}{2^n(n\,!)^2}$$

$$\therefore\quad P_n(x)=\frac{(2n)\,!}{2^n(n\,!)^2}\cdot$$

$$\left[x^n-\frac{n(n-1)}{2(2n-1)}x^{n-2}+\frac{n(n-1)(n-2)(n-3)}{2\cdot 4\,(2n-1)\,(2n-3)}x^{n-4}-\ldots\ldots\right]$$

Rodrigue's Formula (for Legendre's polynomial) :

$$(1-x^2)\frac{d^2v_n}{dx^2}-2x\frac{dv_n}{dx}+n(n+1)v_n=0 \quad \text{where} \quad V_n=\frac{d^nv}{dx^n}$$

V_n is a polynominal of degree n ;

$P_n(x)$ is a constant multiple of V_n ;

$$V_n=\frac{d^nv}{dx^n}=\frac{d^n}{dx^n}(x^2-1)^n$$

$$P_n(x)=cv_n=c\frac{d^n}{dx^n}(x^2-1)^n$$

$$(n+1)P_{n+1}(x)-(2n+1)x\,P_n(x)+nP_{n-1}(x)=0$$

$$x\frac{dP_n}{dx}-\frac{d\,.\,P_{n-1}}{dx}=n\,.\,P_n$$

Orthogonal Property :

$$\int_{-1}^{1} P_m(x)\,P_n(x)\,dx=0 \quad \text{if} \quad m\neq n$$

Gamma Function :

$$\Gamma(n)=\int_0^{\infty} e^{-x}x^{n-1}\,dx \qquad \text{where } n>0$$

$$\Gamma(n+1)=n\Gamma(n)=n\,! \qquad \Gamma(n)=\frac{\Gamma(n+1)}{n}$$

Beta Function :

$$\beta\,(m,n)=\int_0^1 x^{m-1}\,.\,(1-x)^{n-1}dx \qquad \text{for } m>0,\ n>0$$

$$\beta\,(m.\,n)=\beta(n,m)$$

$$\beta\,(m,n)=\frac{\Gamma(m)\,.\,\Gamma(n)}{\Gamma(m+n)}$$

Bessel Function :

$$x\frac{d^2y}{dx^2}+x\frac{dy}{dx}+(x^2-n^2)\,y=0 \ldots\ldots \text{Bessel's eqn. of order } n.$$

$$y=a_0x^n\left\{1-\frac{x^2}{2^2(n+1)}+\frac{x^4}{2^4\,.\,2\,!\,(n+1)(n+2)} \ldots\ldots+\frac{(-1)^rx^{2r}}{2^{2r}\,.\,r\,!\,(n+1)(n+2)\ldots\ldots(n+r)}+\ldots\ldots\right\}$$

Bessel's function of the first kind of nth order is defined as :

$$J_n(x)=\frac{x^n}{2^n\Gamma(n+1)}\left[1-\frac{x^2}{2^2(n+1)}+\frac{x^4}{2^4 . 2!\ (n+1)(n+2)}-\cdots\cdots\right]$$

$$=\sum_{r=0}^{\infty}\frac{(-1)^r}{r!\ \Gamma(n+r+1)}\left(\frac{x}{2}\right)^{n+2r}$$

$$=\frac{x^n}{2^n . n!}\left\{1-\frac{x^2}{2(2n+2)}+\frac{x^4}{2\cdot 4(2n+2)(2n+4)}-\cdots\cdots\right\}$$

which converges for all values of x (real and imaginary), when n is an integer.

$$\text{Thus,}\quad J_n(x)=\sum_{r=0}^{\infty}(-1)^r . \left(\frac{x}{2}\right)^{2r+n}\Big/ r!\ (n+r)$$

If n is not an integer,

$$J_n(x)=\sum_{r=0}^{\infty}(-1)^r\left(\frac{x}{2}\right)^{2r+n}\Big/ r!\ \Gamma(n+r+1)$$

$$J_{-n}(x)=\sum_{r=0}^{\infty}(-1)^r\left(\frac{x}{2}\right)^{2r-n}\Big/ r!\,\Gamma(-n+r+1)$$

$$\therefore\quad J_{-n}(x)=(-1)^n J_n(x)$$

Recurrence Formulas :

$$e^{\frac{x}{2}\left(t-\frac{1}{t}\right)}=\sum_{n=-\infty}^{\infty}J_n(x)\ t^n$$

$$J_{n-1}(x)+J_{n+1}(x)=\frac{2n}{x}\ J_n(x)$$

$$J^1{}_n(x)=\frac{J_{n-1}(x)-J_{n+1}(x)}{2}$$

$$x^n J_{n-1}(x)=\frac{d}{dx}\left\{x^n J_n(x)\right\}$$

$$\frac{d}{dx}\left\{x^{-n}J_n(x)\right\}=-x^{-n}J_{n+1}(x)$$

Bernoullian Trials :

Let p=probability of an event

q=probability of the alternative event

$=1-p$

In n repeated and independent trials, the probability that the event will occur r times is $p^r(1-p)^{n-r}=p^r q^{n-r}$

If r be the no. of occurence with greatest probability, then,

$$\frac{r}{n+1}<p<\frac{r+1}{n+1}$$

Theoretical Binomial Distribution :

Let P=constant probability of success in n independent trials, the probability of occurence of x successes in any order is

$$f(x)={}^n c_x\, P^x\, Q^{n-x}=\frac{n\,!}{(n-x)!\;x!}\,P^x\, Q^{n-x}$$

Moments about origin :

First moments, $\mu_1{}^1=nP$

Second moment, $\mu_2{}^1=nP+n(n-)P^2$

Third moment, $\mu_3{}^1=nP+3n(n-1)P^2+n(n-1)(n-2)P^3$

Central Moments :

First moment, $\mu_1=nP$=Expectation $(x)=E(x)$

Second moment, $\mu_2=\mu_2{}^1-(\mu_1{}^1)^2=nPQ$=variation (x)=var (x).

Standard deviation$=\sqrt{\text{var.}(x)}=\sqrt{nPQ}$

Third moment, $\mu_3=\mu^1{}_3-3\mu_2{}^1\mu_1{}^1+2\mu^1{}_1{}^3$:

Poission's Distribution :

$$f(x)=\frac{c^{-\lambda}}{x\,!}\,\lambda^x\,;$$

The variable x has the probability distribution of N.

The above form may be regarded as a limiting case of a binomial distribution where $n\to\infty$ and $P\to0$, but $nP=\lambda$ remains finite.

Moments about origin :

First moment, $\mu_1{}^1=\lambda$

Second moment, $\mu_2{}^1=\lambda^2+\lambda$

Third moment, $\mu_8{}^1=\lambda^3+3\lambda^2+\lambda$; and so on.

Moments about mean :

Mean for Possion's distribution

$=\mu=\mu_1{}^1=\lambda$

First moment, $\mu_1=0$
Second moments, $\mu_2=\lambda=$ var.
Third moment, $\mu_3=\lambda$
Standard deviation $=\sqrt{\lambda}$

Normal (Gaussian) Distribution :

$$f(x)=\frac{h}{\sqrt{\pi}}\, e^{-h^2(x-a)^2}$$

$$\text{or, } f(x)=\frac{1}{\sigma\sqrt{2\pi}}\, e^{-\frac{(x-\mu)^2}{2\sigma^2}}$$

where $h=\frac{1}{\sigma\sqrt{2}}$

and $-a\leqslant x\leqslant a$.

where $h>0$ and the continuous variable x can take any value between $-a$ and a. $f(x)$ is $+ve$ for all values of x.

$$\int_{-a}^{a} f(x)\, dx=1$$

Standard normal variable,

$r=\frac{x-\mu}{\sigma}$ which has the probability density function

$$\phi(T)=\frac{1}{\sqrt{2\pi}} e^{-\frac{T^2}{2}}$$

Expectation :

$$E(x)=\mu$$

Variance :

$$\text{Var. } (x)=\sigma^2$$

Gamma Distribution :

$$f(x)=\frac{e^{-x}x^{l-1}}{\Gamma(l)}\ ; \text{ where } \Gamma(l)=\int_0^a e^{-x}x^{l-1}\, dx \text{ and}$$

x lies between 0 and x

$$\therefore \int_0^a f(x)\, dx=1$$

$E(x)=l$ var. $(x)=l$

If $x=a$ normal variate with mean μ and standard deviation σ, then gamma variate

$$=\frac{1}{2}\ \frac{(x-\mu)^2}{\sigma^2}$$

x^2-distribution :

$$f(x)=\frac{1}{2^{n/2}\,\Gamma\left(\frac{n}{2}\right)}\,e^{-\frac{x}{2}}\,x^{\left(\frac{n}{2}-1\right)};\quad x>0$$

$$E(x)=n$$

Probabilities :

1. If E is an event, P (not E)$=1-P(E)$.
2. If out of a total of n cases the event E occurs in m cases, the event (not E) will occur in $n-m$ cases.

$$\therefore \quad P(E)=\frac{m}{n} \text{ and } P(\text{not } E)=1-P(E)=1-\frac{m}{n}=\frac{n-m}{n}$$

3. If two events E and F are disjoint (*i.e.*, they never occur together, mutually exclusive),

$$P(E \text{ or } F)=P(E)+P(F)$$

4. **Stochastic Independence :**

$$P\left(\frac{A}{B}\right)=P(A) \text{ and } P\left(\frac{B}{A}\right)=P(B)$$

$$P(A \text{ and } B)=P\left(\frac{A}{B}\right).\,P(B)=P\left(\frac{B}{A}\right).\,P(A)$$

or $$=P(A)\,.\,P(B)=P(B)\,.\,P(A)$$

The occurence or non-occurence of anyone of A and B does not influence the occurence or non-occurence of the other.

Mathematical Theory :

Let A_1 and A_2 be two sets of events. The union of two sets is Ω dedenoted by $A_1 \cup A_2$.

The intersection of these two events is denoted by $A_1 \cap A_2$ Ω denotes sure event and ϕ denotes null event.

Addition of Probabilities :

$$P(A \cup B)=P(A)+P(B)-P(AB)$$

$$P(A \cup B \cup C)=[P(A)+P(B)+P(C)] -[P(AB)+P(AC)+P(BC)]+P(ABC).$$

8

Application of Calculus

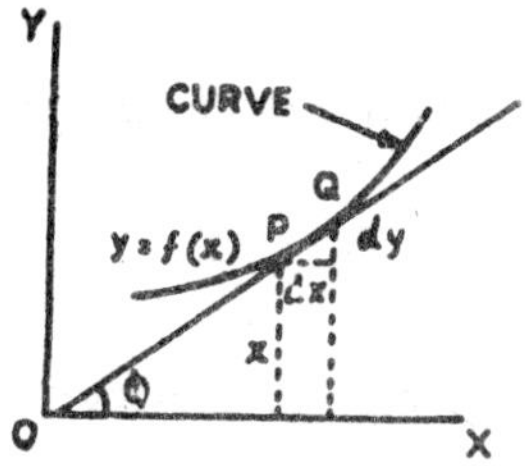

Co-ordinates of $P = x, y$

" " " $Q = (x+dx), (y+dy)$

ϕ = inclination of tangent to curve.

$\frac{dy}{dx} = \tan\phi$ = slope or gradient of curve.

Eqn. of tangent, $y - y_1 = \frac{dy}{dx}(x - x_1)$

Eqn. of normal, $y - y_1 = \frac{-1}{\frac{dy}{dx}}(x - x_1)$

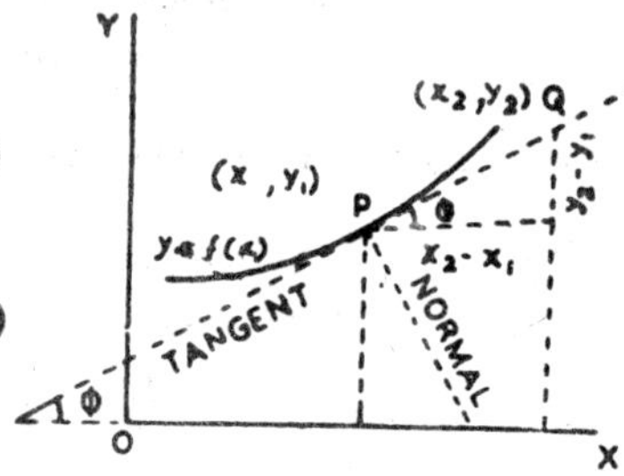

Angle between curves,

$$\theta = \tan^{-1}\frac{\tan\phi_1 - \tan\phi_2}{1 + \tan\phi_1\tan\phi_2}$$

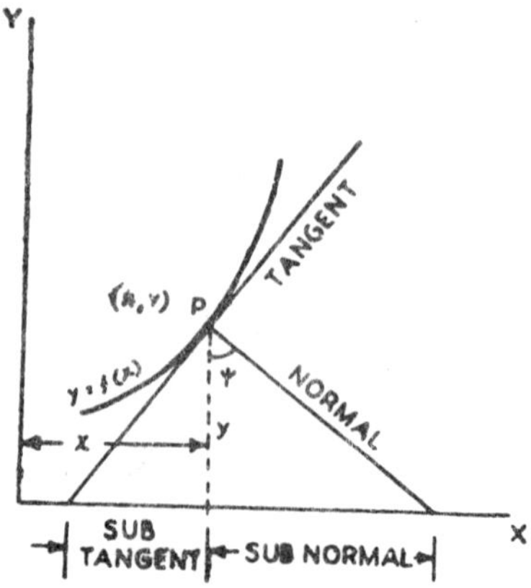

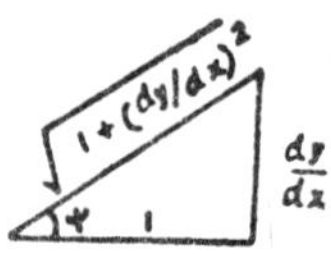

Length of tangent $= \dfrac{y\sqrt{1+\left(\dfrac{dy}{dx}\right)^2}}{\dfrac{dy}{dx}}$

Length of normal $= y\sqrt{1+\left(\dfrac{dy}{dx}\right)^2}$

Length of subtangent

$$= \frac{y}{\dfrac{dy}{dx}}$$

Length of sub normal

$$= y\,\frac{dy}{dx}$$

Differential co-efficient of an arc :

(*a*) Cartesian equation

$$\left(\frac{ds}{dx}\right)^2 = 1 + \left(\frac{dy}{dx}\right)^2$$

$$\left(\frac{ds}{dy}\right)^2 = 1 + \left(\frac{dx}{dy}\right)^2$$

$$\cos\psi = \frac{dx}{ds}\ ;\ \sin\psi = \frac{dy}{ds}$$

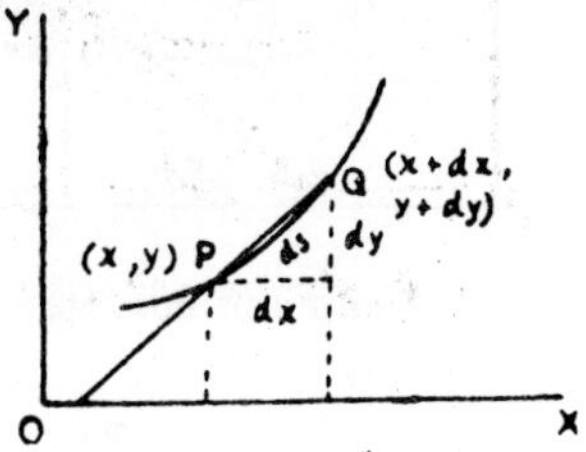

(*b*) Polar equation

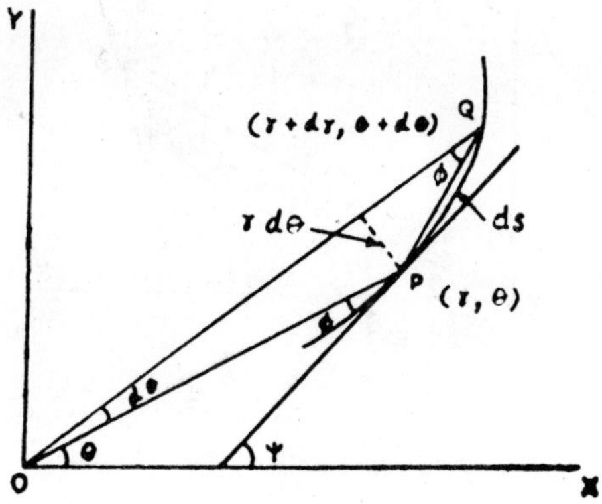

$$\left(\frac{ds}{d\theta}\right)^2 = r^2 + \left(\frac{dr}{d\theta}\right)^2$$

$$\left(\frac{ds}{dr}\right)^2 = 1 + \left(\gamma\frac{d\theta}{dr}\right)^2$$

$$\cos\phi = \frac{dr}{ds} \;;\; \sin\phi = r\frac{d\theta}{ds}$$

Radius of curvature at (x, y) of curve given by Cartesian equation y=f(x)

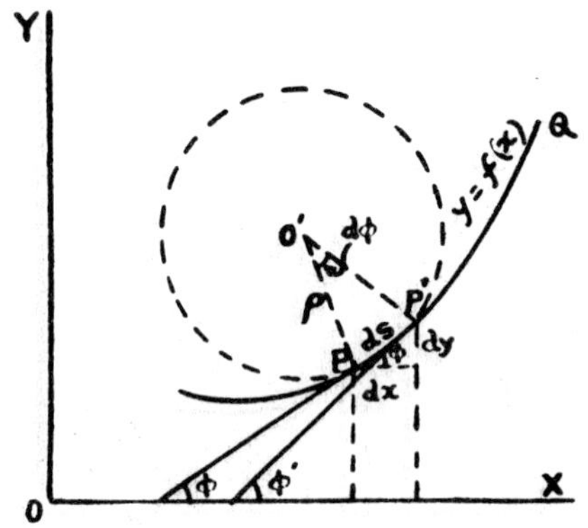

$$\rho = \frac{(1+y_1^2)^{3/2}}{y^2}$$

Radius of curvature at θ of curve given by parametric equation

$$x = f(\theta),\; y = \phi(\theta)$$

$$\rho = \frac{(x'^2+y'^2)^{3/2}}{x'y''-y'x''}$$

Radius of curvature of curve given by pedal equation p=f(r)

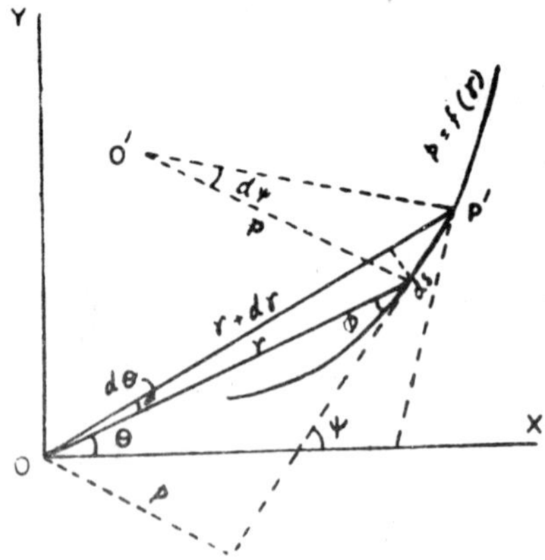

$$\rho = \frac{ds}{d\psi} = r\,\frac{dr}{dp}$$

Radius of curvature of curve given by polar equation $r=f(\theta)$

$$\rho=\frac{(r^2+r_1^2)^{3/2}}{r^2+2r_1^2-rr_2}$$

Area bounded by curve given by polar equation

$$\text{Area, A} =\frac{1}{2}\int_{\theta_1}^{\theta_2} r^2\, d\theta$$

$$=\frac{1}{2}\int_{\theta_1}^{\theta_2} [f(\theta)]^2\, d\theta$$

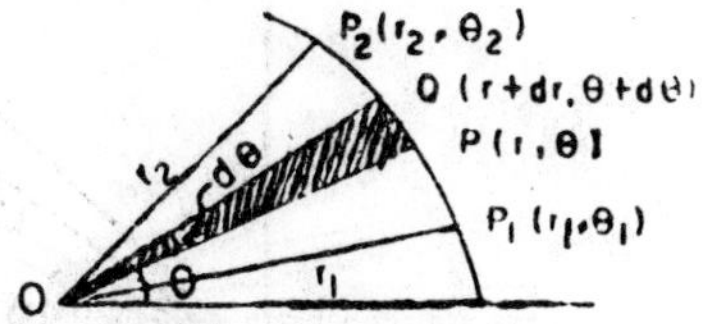

Length of curve :

Cartesian equation :

From $x=a$ to $x=b$,

$$S=\int_a^b\sqrt{1+\left(\frac{dy}{dx}\right)^2}\,.\,dx$$

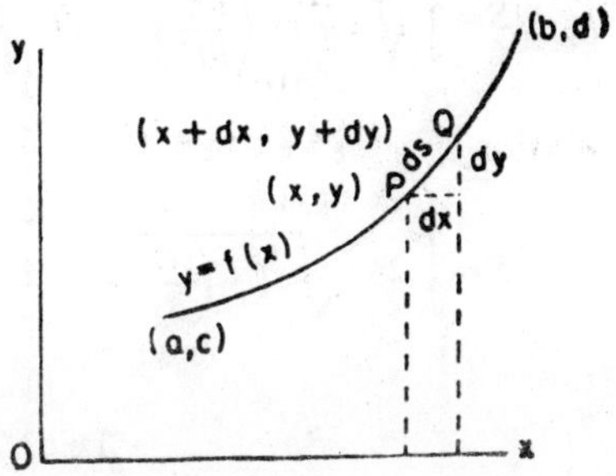

From $y=c$ to $y=d$,

$$S=\int_c^d\sqrt{1+\left(\frac{dx}{dy}\right)^2}\,.\,dy$$

Parametric equation :

$$S=\int_{t_1}^{t_2}\sqrt{\left(\frac{dx}{dt}\right)^2+\left(\frac{dy}{dt}\right)^2}\,.\,dt$$

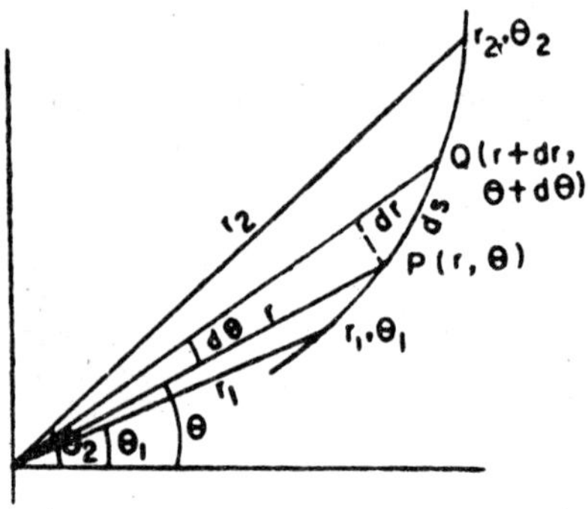

Polar equation :

$$S=\int_{\theta_1}^{\theta_2}\sqrt{r^2+\left(\frac{dr}{d\theta}\right)^2}\;.\;d\theta\text{...from } \theta_1 \text{ to } \theta_2$$

$$S=\int_{r_1}^{r_2}\sqrt{1+r^2\left(\frac{d\theta}{dr}\right)^2}\;.\;dr\text{...from } r_1 \text{ to } r_2$$

Surface of revolution :

$$S=\int_{c}^{d}2\pi y\sqrt{1+\left(\frac{dx}{dy}\right)^2}\;.\;dy$$

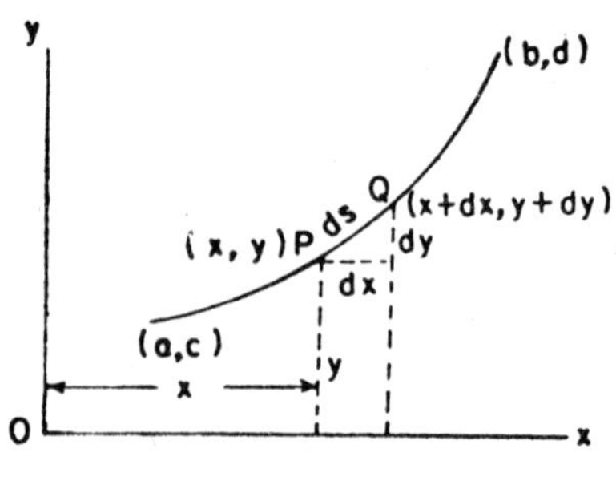

$$S=\int_a^b 2\pi x \sqrt{1+\left(\frac{dy}{dx}\right)^2}\ .\ dx$$

$$S=2\pi\int r \sin\theta\ .\ \sqrt{r^2+\left(\frac{dr}{d\theta}\right)^2}\ .\ d\theta,$$

when the curve is given by polar equation $r=f(\theta)$.

Simple Harmonic Motion :

$$\frac{d^2x}{dt^2}=-\omega^2 x$$

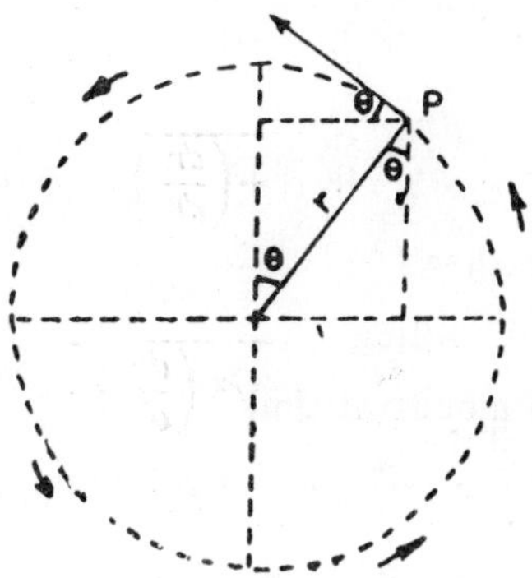

Simple Pendulum :

$$\frac{d^2s}{dt^2}=-\frac{g}{l}\ .\ S$$

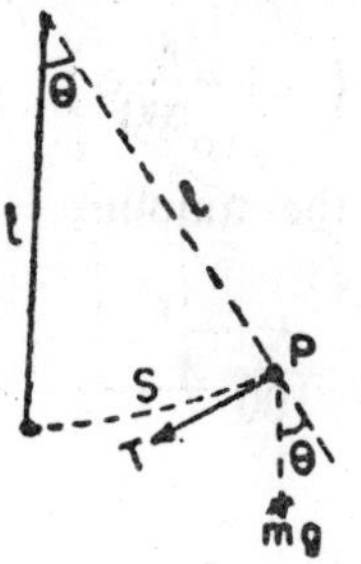

Motion of a elastic string or helical spring :

Let K=stiffness of the string and

W=weight suspended from the string.

By Newtons second law of motion,

force=mass×acceleration

for x displacement of mass, force applied=Kx,

$$\text{mass}=\frac{W}{g} \text{ and acceleration}=\frac{d^2x}{dt^2}$$

$\therefore$ Eqn. of motion is $\frac{W}{g} \cdot \frac{d^2x}{dt^2}=-Kx$

$$i.e., \quad \frac{d^2x}{dt^2}+\frac{Kg}{W}x=0$$

Cooling of hot bodies t

Newton's law of cooling :

Let θ=temp. of a body at time t

α=const. temp. of the surrounding medium. Then, rate of change of temperature

$$\frac{d\theta}{dt}=-K(\theta-\alpha), \ K \text{ being a constant.}$$

Compound interest :

For an amount A_0 invested at $K\%$ interest, the amount at the end of y years (after being compounded every year) is given by

$$A_t=A_0\left(1+\frac{K}{100}\right)t$$

The rate of change in the amount A at any time is proportional to the amount at the time ;

$$i.e., \quad \frac{dA}{dy}=\frac{K}{100}A$$

Radio-active decay :

The rate of decay (*i.e.*, loss of weight) at any time is proportional to the amount of the element remaining at the time.

Let x gms be the amount remaining at time t ;

Then, $\frac{dx}{dt}=-Kx$; where K is a constant.

Growth of population :

Let P=population of a city at time t : Then, rate of growth of population

$$\frac{dP}{dt}=KP$$

Current in A.C. circuits :

For circuits with inductance and alternating e.m.f.,

$$L\frac{di}{dt}=E\sin\omega t.$$

For circuits with resistance, inductance and constant e.m.f.,

$$Ri+L\frac{di}{dt}=E$$

For circuits with resistance, inductance and alternating e.m.f.

$$L\frac{di}{dt}+Ri=E\sin\omega t$$

For circuits with resistance, inductance and capacitance,

$$L\frac{d^2q}{dt^2}+R\frac{dq}{dt}+\frac{q}{c}=E$$

Potential drop in inductance

$$=L\frac{di}{dt}=L\frac{d^2q}{dt^2} \qquad \left[\because \quad i=\frac{dq}{dt}\right]$$

Potential drop in resistance

$$=Ri=R\frac{dq}{dt}$$

Potential drop in capacitor

$$=\frac{q}{c}$$

9

Statics

(SCIENCE OF EQUILIBRIUM)

Parallelogram of forces :

Resultant, $R=\sqrt{P^2+Q^2+2PQ\cos\alpha}$
when $\alpha=0$, P and Q will act in the same direction and $R=P+Q$;

when $\alpha=180°$, P and Q will act in opposite direction and $R=P-Q$

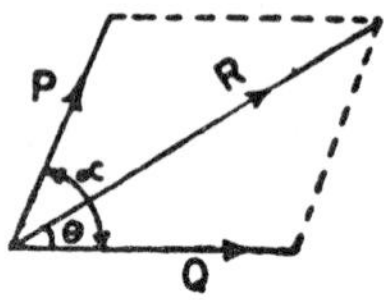

Refer. to the above fig.,

$$\tan\theta=\frac{Q\sin\alpha}{P+Q\cos\alpha}$$

$$\text{or,}\quad \theta=\tan^{-1}\left(\frac{\theta\sin\alpha}{P+Q\cos\alpha}\right)$$

Theorem of resolved parts

The algebraic sum of the resolved parts of any two concurrent forces in any direction, is equal to the resolved part of their resultant in the same direction.

Triangle of forces

If three forces acting at a point be represented in magnitude, direction and sense (but not in position) by the three sides of a triangle, then the forces are in equilibrium and vice-versa.

Polygon of forces

If any number of forces acting at a point be represented in magnitude, direction and sense by the sides of a polygon, taken in order, then they shall be in equilibrium.

Lami's theorem

When three forces are in equilibrium, each one is proportional to the sine of the angle between the other two.

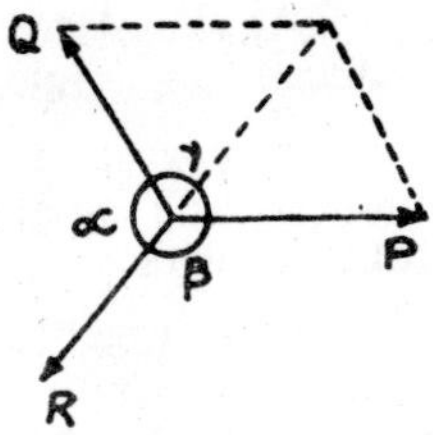

$$\frac{P}{\sin \alpha}=\frac{Q}{\sin \beta}=\frac{R}{\sin \gamma};$$

Moment

The moment of a force about a particular point is the product of the force and the perpendicular distance of the line of action of the force from that point. When the moment tends to produce clockwise rotation it is considered positive and anti-clockwise moment is taken to be negative.

Varignon's theorem

The algebraic sum of the moments of any two coplanar forces about a point is equal to the moment of their resultant about that point.

Couple

Two equal and unlike parallel forces constitute a couple.

The algebraic sum of the moments of the two forces forming a couple about a point is equal to the moment of the couple.

The moment of a couple can never be zero.

If the algebraic sum of the moments of a pair of coplanar forces about a point remains constant, then they constitute a couple.

Two co-planar couples acting upon a rigid body having equal and opposite moments balance each other.

The effect of a couple is not changed when transferred to a parallel plane, provided the moment of the couple is unchanged in magnitude and sense.

A couple and a force acting in the same plane are equivalent to a single force, equal and parallel to the given single force.

If three forces acting in the same plane upon a rigid body be represented in magnitude, direction, position, and sense, by the three sides of a triangle taken in order, they are equivalent to a couple whose moment is twice the area of the triangle.

Coplanar forces

Any system of coplanar forces acting on a rigid body can be reduced to a single force or a single couple unless it is in equilibrium.

Any system of coplanar forces acting on a rigid body can be reduced to a single force acting at any arbitrary point in the same plane together with a couple.

If the algebraic sum of the moments of a system of forces (acting in a plane upon a rigid body) about each of three points in the plane (not bying in the same line) vanish separately, the system of the forces is in equilibrium.

If a rigid body is kept in equilibrium by the action of three coplanar forces, the forces must be parallel or they meet at a point.

The algebraic sum of a system of coplanar forces about each of three non-collinear points is M and that is not zero, the system of forces is reduced to a couple of moment M.

Conditions of equilibrium

(1) Algebraic sum of all horizontal forces acting on a structure or a member is zero *i.e.*, $\Sigma H=0$

(2) Algebraic sum of all vertical forces acting on a structure or a member is zero, *i.e.*, $\Sigma V=0$.

(3) Algebraic sum of all moments about a point in a stucture is zero *i.c.*, $\Sigma M=0$.

Centre of gravity

A body has one and only one centre of gravity.

(1) The centre of gravity of a thin uniform rod lies at its centre.

(2) The centre of gravity of a circle is at its centre.

(3) The centre of gravity of rectangle or a parallelogram is at the intersection of diagonals or what is the same thing, at the intersection of lines joining the mid-points of opposite sides.

(4) The centre of gravity of a triangle is at the intersection of the medians.

(5) The centre of gravity of a flat thin circular segment is on the axis of symmetry at 2/5 of its height above the base.

(6) The centre of gravity of a uniform triangular lamina is the same as that of three equal particles placed at the vertices of the triangle.

(7) The centre of gravity of a sphere is at its centre.

(8) The centre of gravity of a hemi-sphere is on the axis of symmetry 3/8 the radius from the centre of sphere.

(9) The centre of gravity of a cylinder or prism is in the middle of the straight line connecting the centres of the bases of the cylinder or prism.

(10) The centre of gravity of a pyramid or cone is on the straight line connecting the centre of the base and the apex at 1/4 the distance above the base.

Centre of gravity of area of plane figures :

The area A is supposed to be composed of several small elementary areas, a_1, a_2, etc., their distances from Y-axis are x_1, x_2, etc. and from X-axis are y_1, y_2, etc.

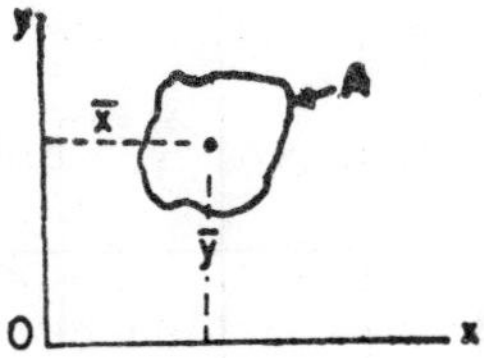

Then, $\bar{x}=\dfrac{\Sigma ax}{A}$

and $\quad \bar{y}=\dfrac{\Sigma ay}{A}$

Second moments of area of plane figures :

Second moment of area with respect to an axis in its plane,

$$I_x=\int y^2\,dA$$
$$I_y=\int x^2\,dA$$

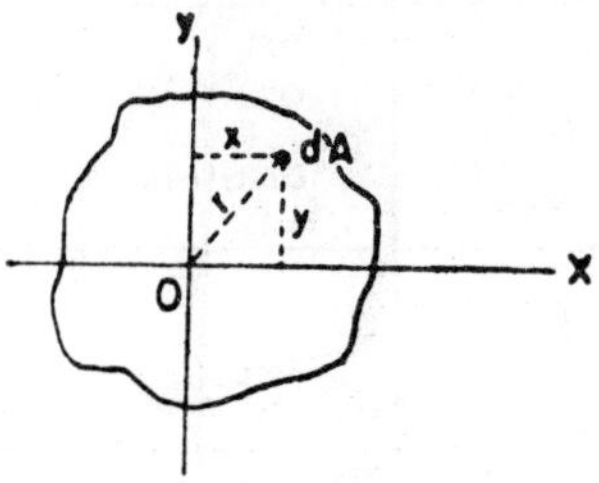

Polar second moment of area,

$$I_p=\int r^2 dA=\int(x^2+y^2)\,dA$$
$$=\int x^2\,dA+\int y^2\,dA$$
$$=I_x+I_y$$

The product of inertia,

$I_x'=\int xy\,dA$ may be positive, or negative or may vanish.

Parallel axis theorem :

$$I_x'=I_x+Ab^2$$

$$I_y'=I_y+Aa^2$$

$$I_p'=I_p+Ad^2$$

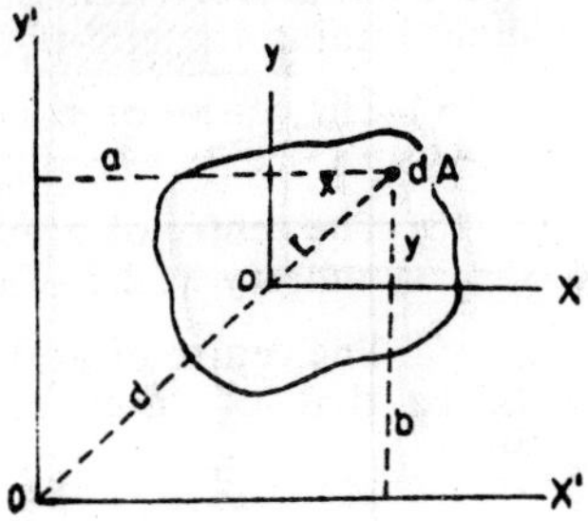

where A=area of the plane figure
and dA=elementary area.

Section	Diagram	Sectional Area	Distances centroid from outer edges	Moments of Inertia			
				I_x	I_y	I'_x	I'_y
Square	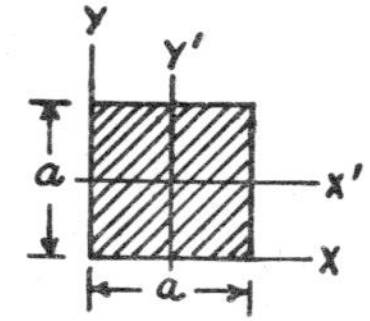	a^2	$\frac{a}{2}, \frac{a}{2}$	$\frac{a^4}{3}$	$\frac{a^4}{3}$	$\frac{a^4}{12}$	$\frac{a^4}{12}$
Hollow Square	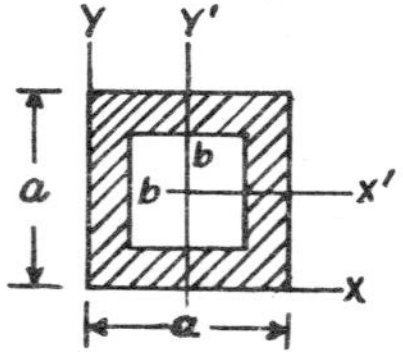	a^2-b^2	$\frac{a}{2}, \frac{a}{2}$	$\frac{a^4}{3}-\frac{b^4}{12}-\frac{a^2b^2}{4}$	$\frac{a^4}{3}-\frac{b^4}{12}-\frac{a^2b^2}{4}$	$\frac{a^4-b^4}{12}$	$\frac{a^4-b^4}{12}$
Rectangle	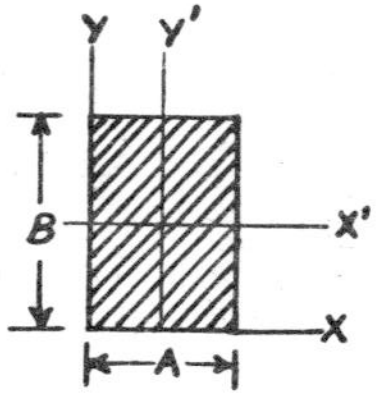	AB	$\frac{A}{2}, \frac{B}{2}$	$\frac{1}{3}AB^3$	$\frac{1}{3}A^3B$	$\frac{1}{12}AB^3$	$\frac{1}{12}A^3B$

Section	Diagram	Sectional Area	Distances of centroid from outer edges	Moments of Inertia			
				I_x	I_y	I'_x	I'_y
Hollow Rectangular	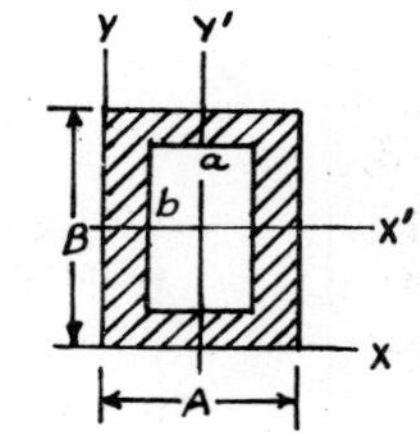	$AB-ab$	$\frac{A}{2}, \frac{B}{2}$	$\frac{AB^3}{2}-\frac{ab}{4}\left(\frac{b^2}{3}-B_2\right)$	$\frac{BA^3}{3}-\frac{ab}{4}\left(\frac{a^2}{3}-A^2\right)$	$\frac{1}{12}(AB^3-ab^3)$	$\frac{1}{12}(A^3B-a^3b)$
Circular	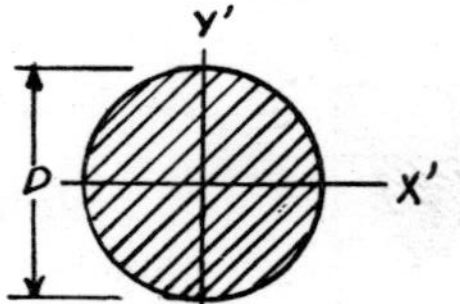	$\frac{\pi D^2}{4}$	$\frac{D}{2}, \frac{D}{2}$			$\frac{\pi D^4}{64}$	$\frac{\pi D^4}{64}$
Hollow Circular	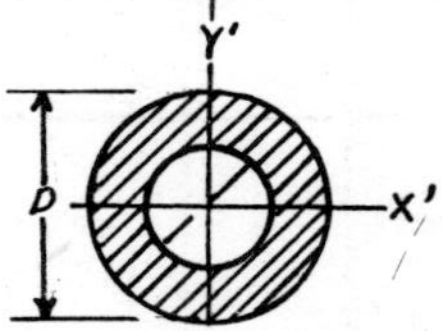	$\frac{\pi}{4}(D^2-d^2)$	$\frac{D}{2}, \frac{D}{2}$			$\frac{\pi}{64}(D^4-d^4)$	$\frac{\pi}{64}(D^4-d^4)$

Section	Diagram	Sectional Area	Distances of centroid from outer edges	Moments of Inertia			
				I_x	I_y	I'_x	I'_y
Triangle	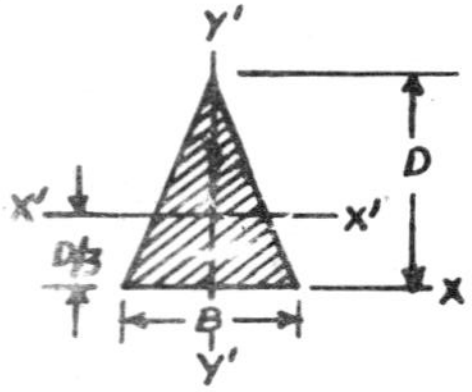	$\frac{BD}{2}$	$\frac{B}{2}$ from each side and $\frac{D}{3}$ from base.	$\frac{BD^3}{12}$		$\frac{BD^3}{36}$	$\frac{B^3D}{36}$
Ellipse	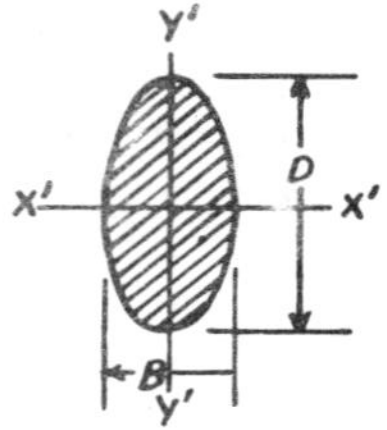	$\frac{\pi BD}{4}$	$\frac{B}{2}, \frac{D}{2}$			$\frac{BD^3}{64}$	$\frac{B^3D}{64}$
Ellipse	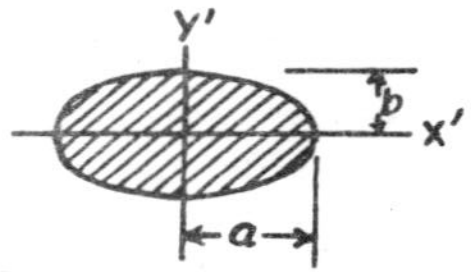	πab	a, b			$\frac{\pi}{4} ab^3$	$\frac{\pi}{4} a^3b$

Section	Diagram	Sectional Area	Distances of centroid from outer edges	Moments of Inertia			
				I_x	I_y	I'_x	I'_y
Parabola	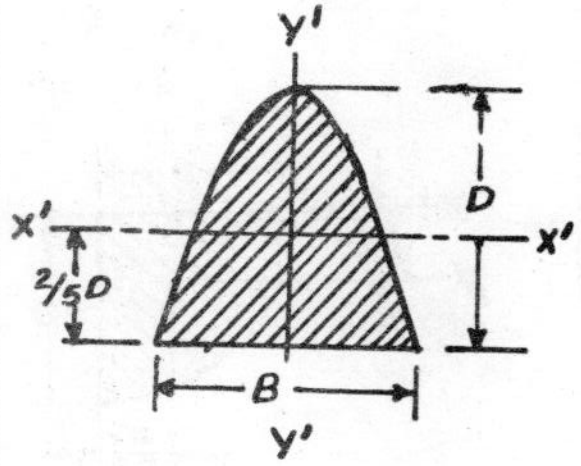	$\frac{2}{3}$BD	$\frac{B}{2}$ from each side and (2/5)D from base.			$\frac{8}{175}BD^3$	$\frac{1}{30}B^3D$
Semicircle	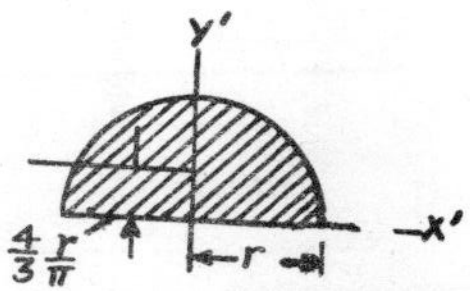	$\frac{\pi r^2}{2}$	r from each side and $\frac{4}{3}\frac{r}{\pi}$ from base			$\frac{\pi r^4}{8}$	$\frac{\pi r^4}{8}$

Section	Diagram	Sectional Area	Distances of centroid from outer edges	Moments of Inertia			
				I_x	I_y	I'_x	I'_y
Quadrant of a Circle	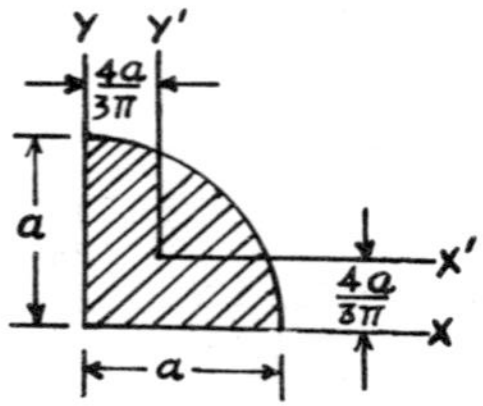	$\frac{\pi a^2}{4}$	$\frac{4a}{3\pi}$, $\frac{4a}{3\pi}$	$\frac{\pi a^4}{16}$	$\frac{\pi a^4}{16}$	$.054\ a^4$	$.054\ a^4$
Quadrant of a Ellipse	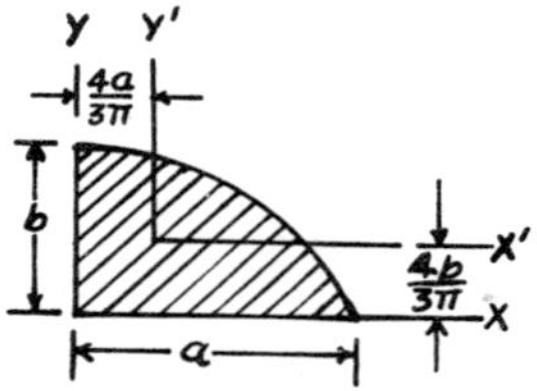	$\frac{\pi ab^2}{4}$	$\frac{4a}{3\pi}$, $\frac{4b}{3\pi}$	$\frac{\pi ab^3}{16}$	$\frac{\pi a^3 b}{16}$	$.054\ ab^3$	$.054\ a^3b$

Section	Sectional Area A	Distance y	Moment of Inertia about xx I_{xx}	Radius of gyration K
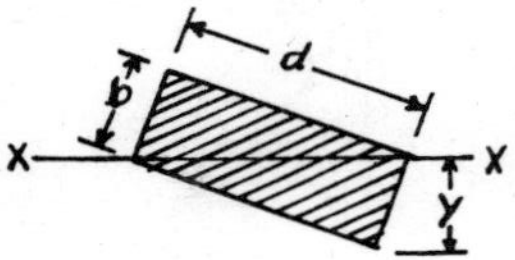	bd	$\dfrac{bd}{\sqrt{b^2+d^2}}$	$\dfrac{b^3d^3}{6\,(b^2+d^2)}$	$\dfrac{bd}{\sqrt{6\,(b^2+d^2)}}$
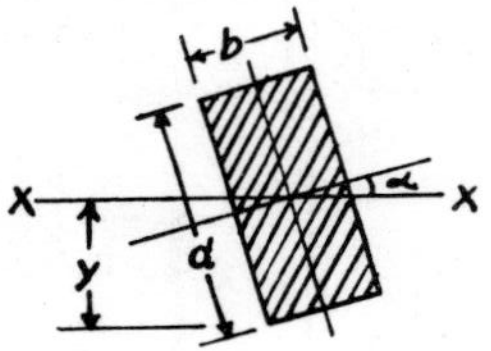	bd	$\dfrac{b \sin \alpha + d \cos \alpha}{2}$	$\dfrac{bd(b^2 \sin^2 \alpha + d^2 \cos^2 \alpha)}{12}$	$\sqrt{\dfrac{b^2 \sin^2 \alpha + d^2 \cos^2 \alpha}{12}}$
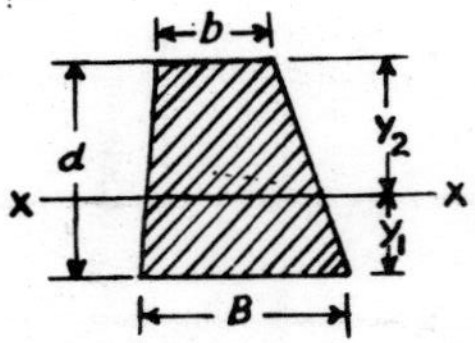	$\dfrac{d}{2}(\mathrm{B}+b)$	$y_1 = \dfrac{d(\mathrm{B}+2b)}{3(\mathrm{B}+b)}$ $y_2 = \dfrac{d(2\mathrm{B}+b)}{3(\mathrm{B}+b)}$	$\dfrac{d^3\,(\mathrm{B}^2+4\mathrm{B}b+b^2)}{36\,(\mathrm{B}+b)}$	$\sqrt{\dfrac{I_{xx}}{\mathrm{A}}}$

Section	Sectional Area A	Distances y	Moment of Inertiu xx I_{xx}	Radius of gyration K
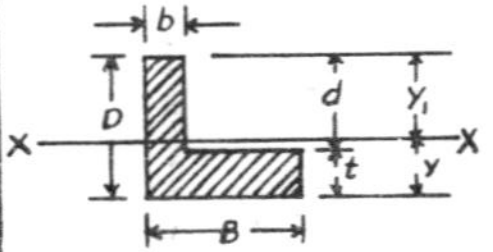	$A = Bt + bd$	$y_1 = \frac{D^2b + t^2(B-b)}{2(Bt+bd)}$ $y = D - y_1$	$\frac{by^3 + By_1^3 - (B-b)(y_1-t)^3}{3}$	$\sqrt{\frac{I_{xx}}{A}}$
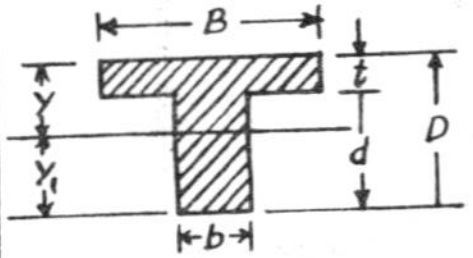	$A = Bt + bd$	$y_1 = \frac{D^2b + t^2(B-b)}{2(Bt+bd)}$ $y = D - y_1$	$\frac{by^3 + By_1^3 - (B-b)(y_1-t)^3}{3}$	$\sqrt{\frac{I_{xx}}{A}}$
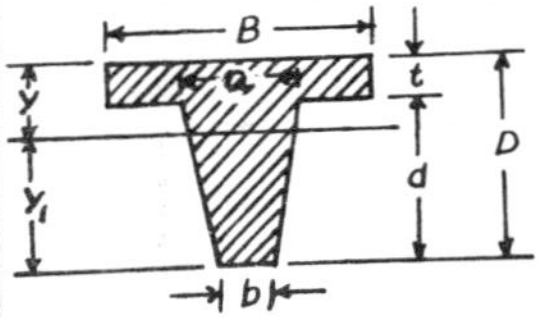	$A = Bt + \frac{d(a+b)}{2}$	$y_1 = \frac{3Bt^2 + 3bd(D+t) + d(a-b)(d+3t)}{6\{2Bt + d(a+b)\}}$ $y = D - y_1$	$\frac{4Bt^3 + d(3b+a)}{12} - A(y_1-t)^2$	$\sqrt{\frac{I_{xx}}{A}}$

Type of loading	Maximum moment	Shear Force Diagram	Bending Moment Diagram	Maximum Deflection
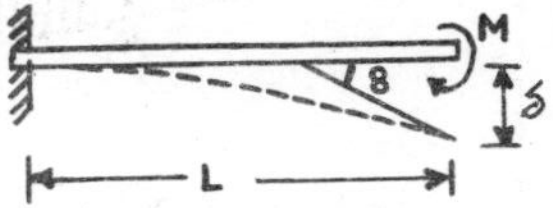	M (−ve)		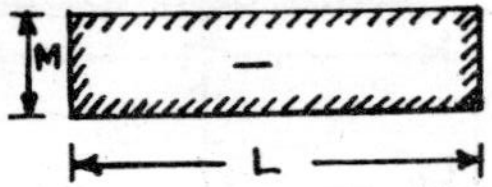	$\frac{ML^2}{2EI}$
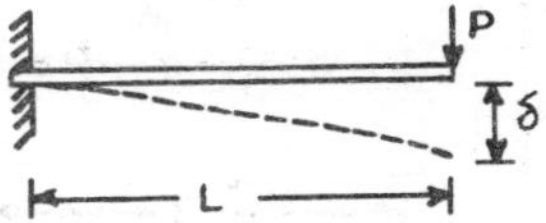	PL (−ve)	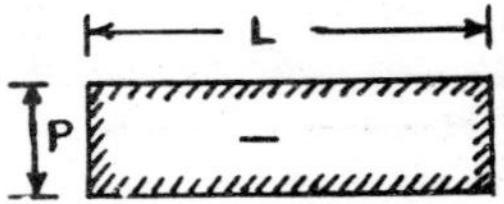	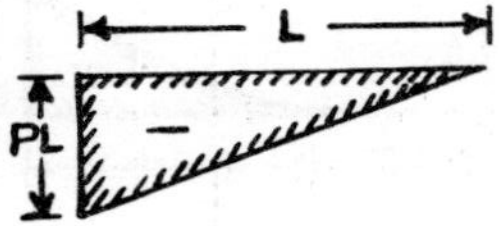	$\frac{PL^3}{3EI}$
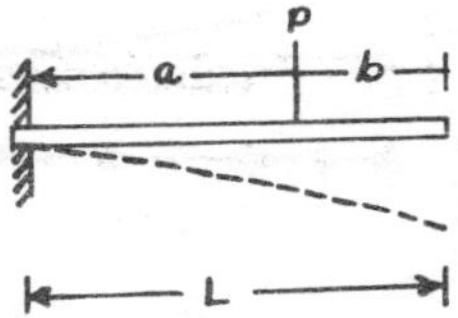	Pa (−ve)	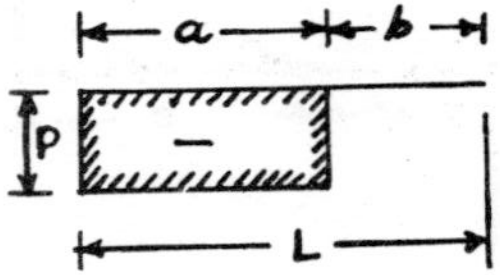	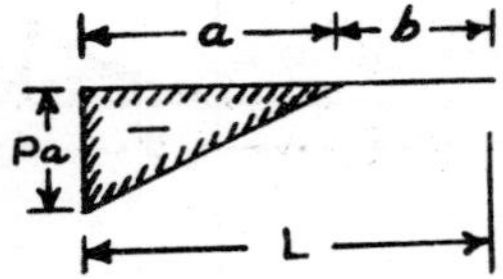	$\frac{Pa^2}{6EI}(3L-a)$
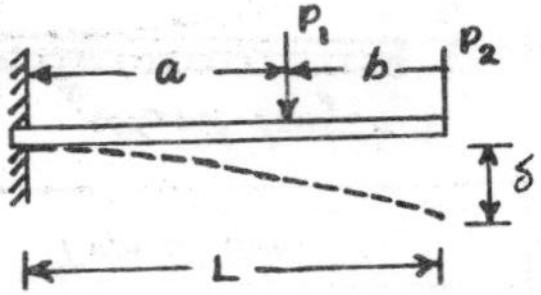	$(P_1+P_2)a + P_2b$ (−ve)	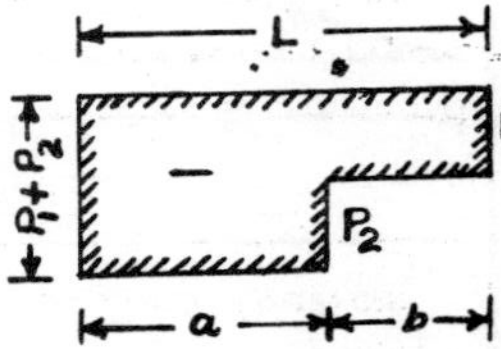	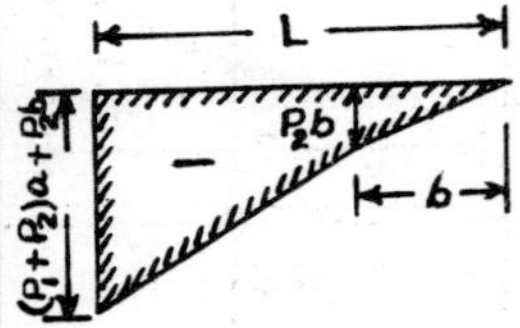	$\frac{P_1a^2(3L-a)+2P_2L^3}{6EI}$

Type of loading	Maximum moment	Shear Force Diagram	B.M. Diagram	Maximum Deflection
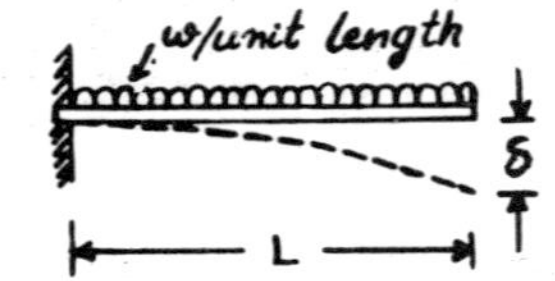	$\frac{wL^2}{2}$ (– ve)	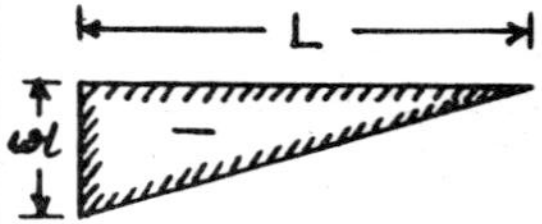	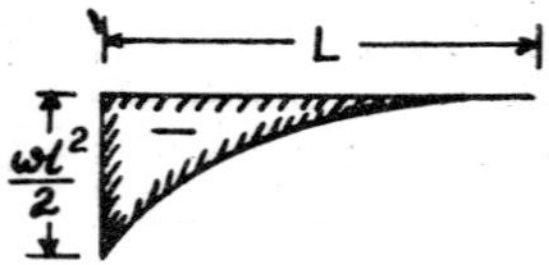	$\frac{wL^4}{8EI}$
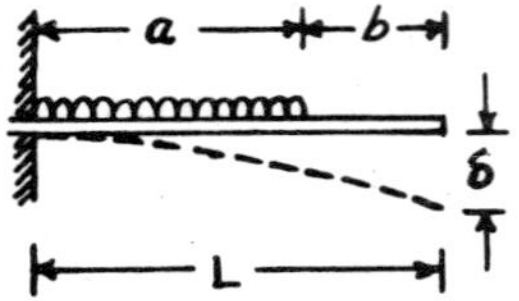	$\frac{wa^2}{2}$ (– ve)	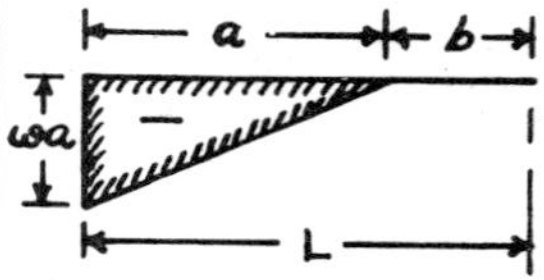	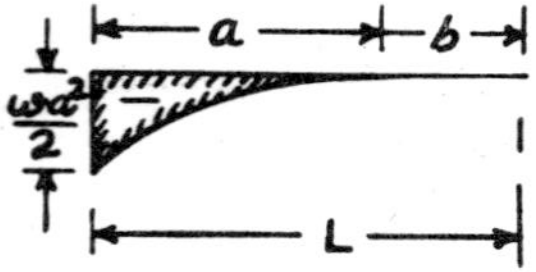	$\frac{wa^4}{8EI} + \frac{wa^3}{6EI}.b$
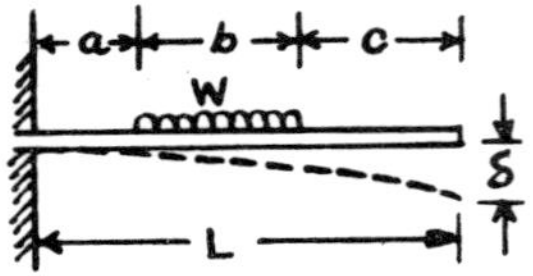	$W\left(a+\frac{b}{2}\right)$ (– ve)	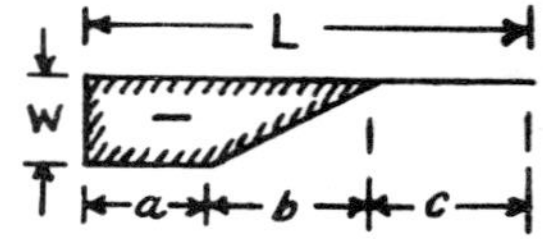	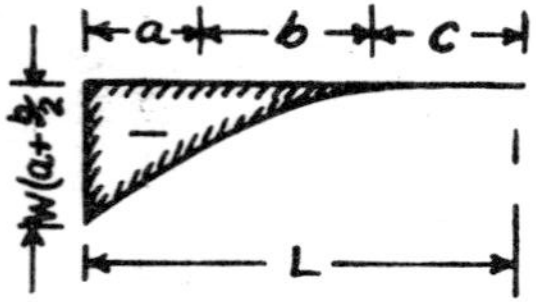	$\frac{W}{24EI}\{4L(3a^2+3ab+b^2) - 4a^3-6a^2b-4ab^2-b^3\}$
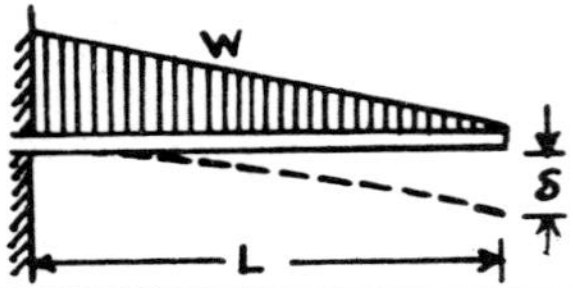	$\frac{WL}{3}$ (– ve)	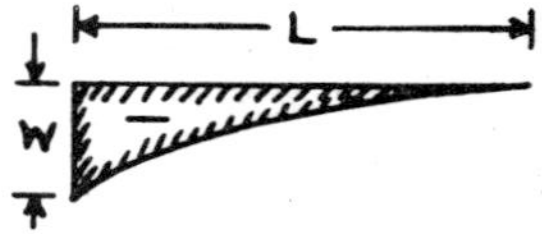	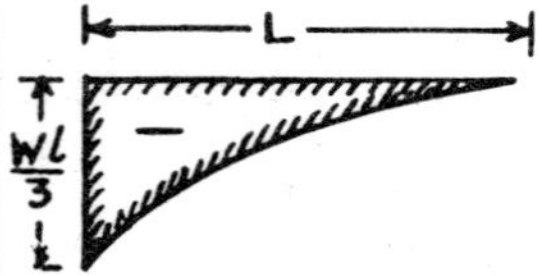	$\frac{WL^3}{15EI}$

Type of loading	*Maximum moment*	*Shear Force Diagram*	*B.M.Diagram*	*Maximum Deflection*
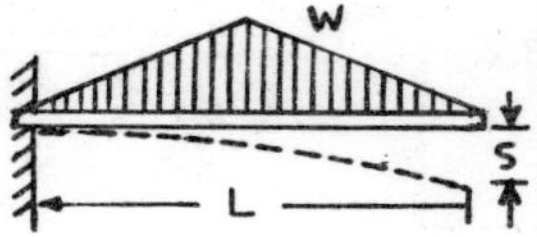	$\frac{WL}{2}$ (−ve)	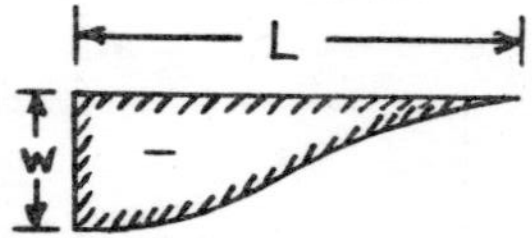	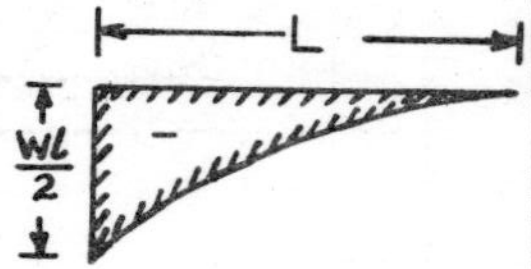	$\frac{11}{96} \cdot \frac{WL^3}{EI}$
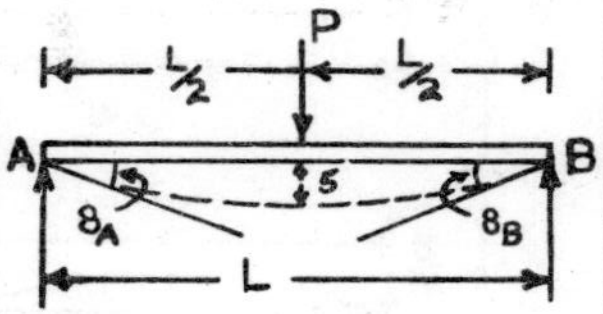	$\frac{PL}{4}$ (+ve)	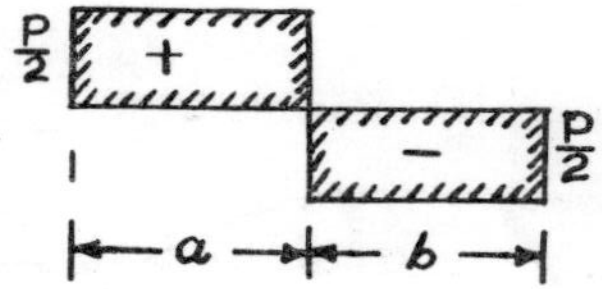	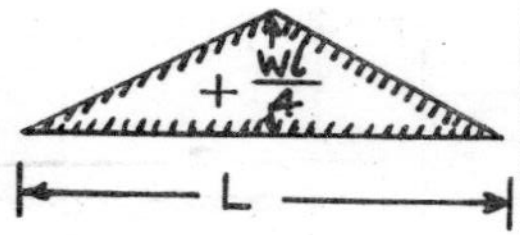	$\frac{WL^3}{48EI}$
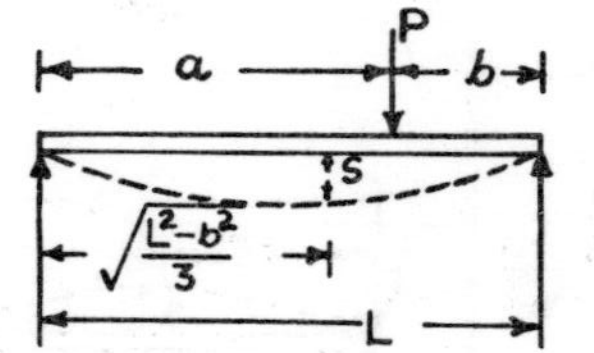	$\frac{Pab}{L}$ (+ve)	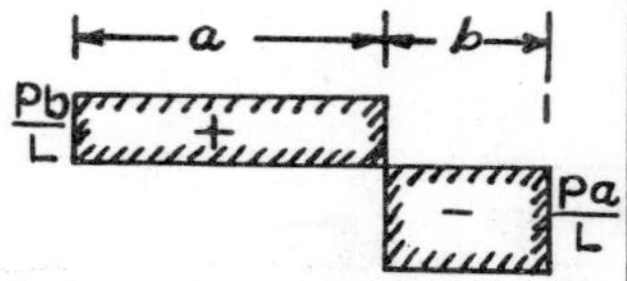	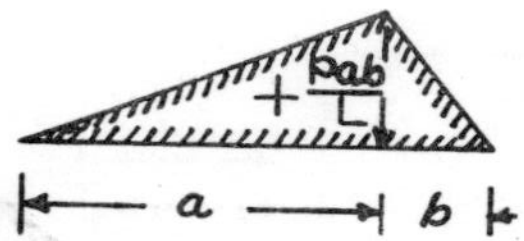	$\frac{Pb}{48EI}(3L^2-4b^2)$ when $a>b$
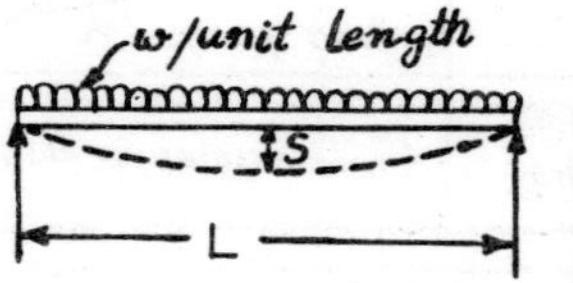	$\frac{wL^2}{8}$ (+ve)	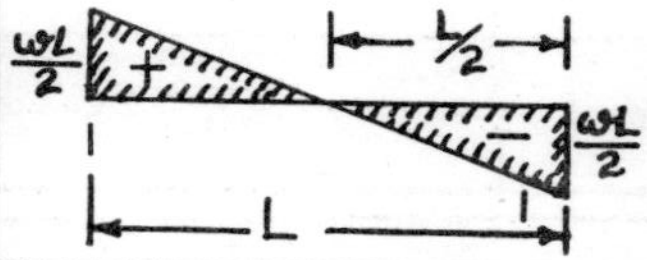	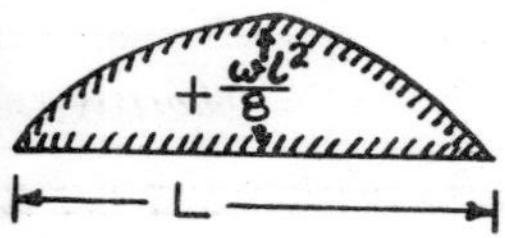	$\frac{5}{384} \cdot \frac{wL^4}{EI}$

Type of loading	*Maximum moment*	*Shear Force Diagram*	*B.M. Diagram*	*Maximum Deflection*
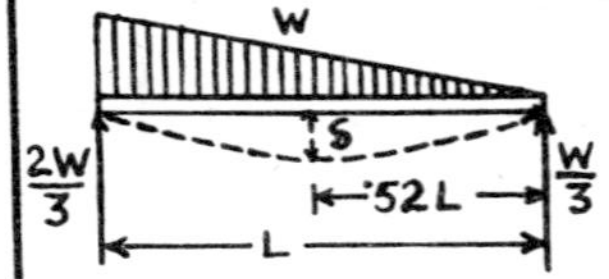	$\frac{2WL}{9\sqrt{3}}$ (+ve)		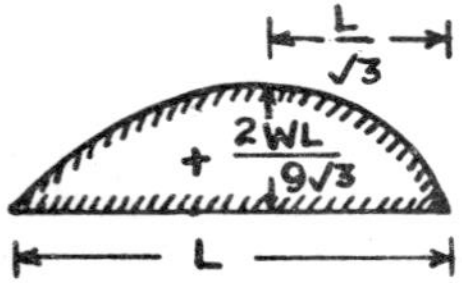	$0{\cdot}013\ \frac{WL^3}{EI}$
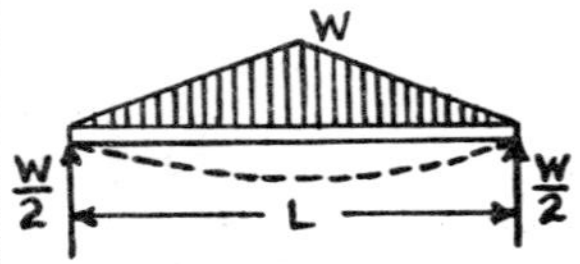	$\frac{WL}{6}$ (+ve)		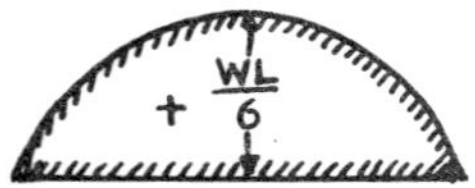	$\frac{WL^3}{60EI}$
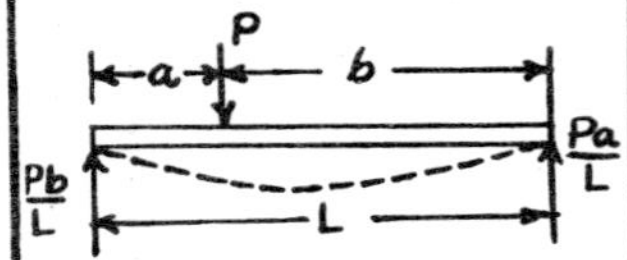	$\frac{Pab}{L}$ (+ve)	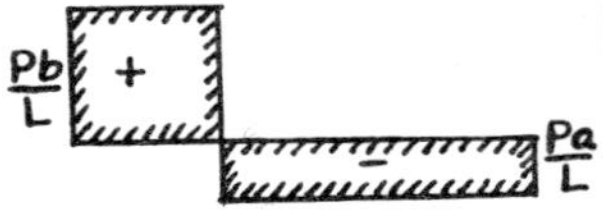	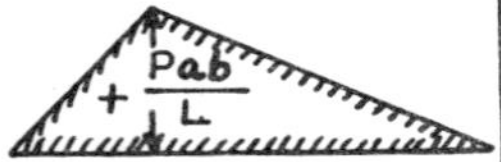	$\frac{Pab}{27EIL}(a+2b)\sqrt{[3a(a+2b)]}$
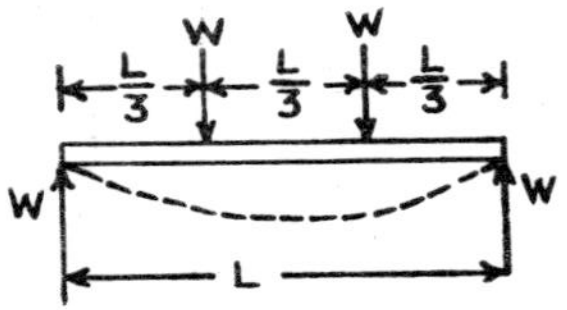	$\frac{WL}{3}$ (+ve)	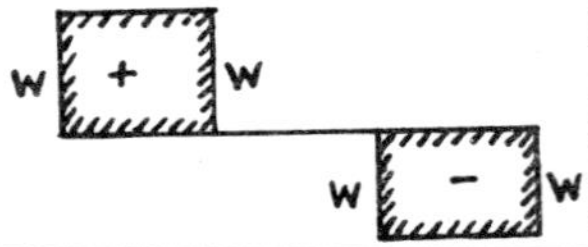	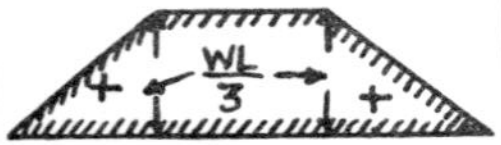	$\frac{23}{648}\cdot\frac{WL^3}{EI}$ *i.e.*, $0.0355\ \frac{WL^3}{EI}$

Type of loading	Maximum moment	Shear Force Diagram	B.M. Diagram	Maximum Deflection
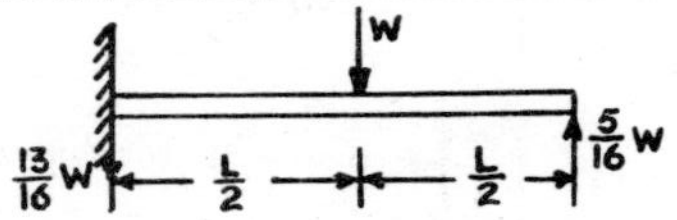	$\frac{3}{16}$WL (+ve)	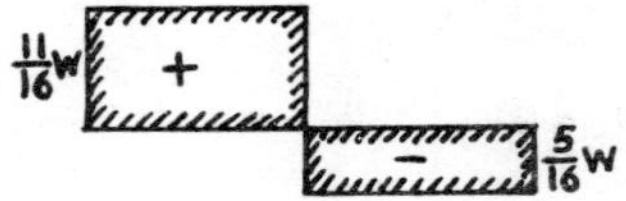	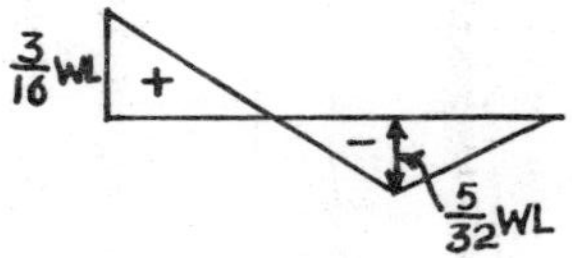	$0.0093\frac{WL^3}{EI}$
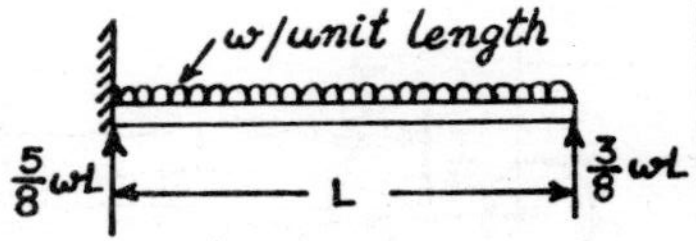	$\frac{wL^2}{8}$ (+ve)		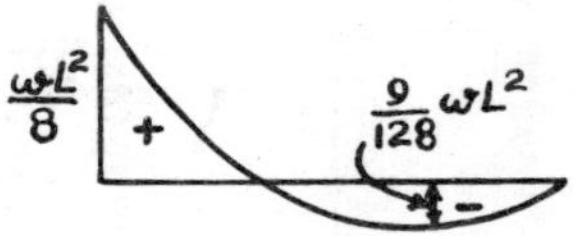	$0.0054\frac{wL^4}{EI}$
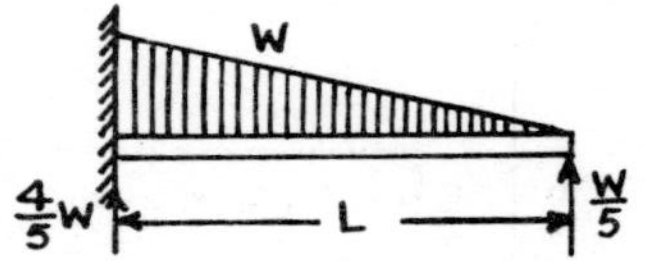	0.133WL (+ve)		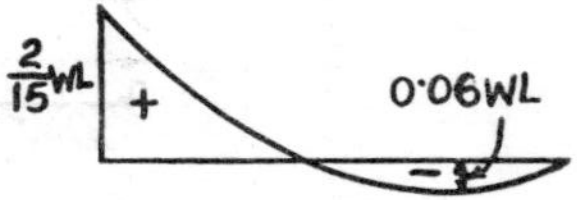	$0.0048\frac{WL^3}{EI}$
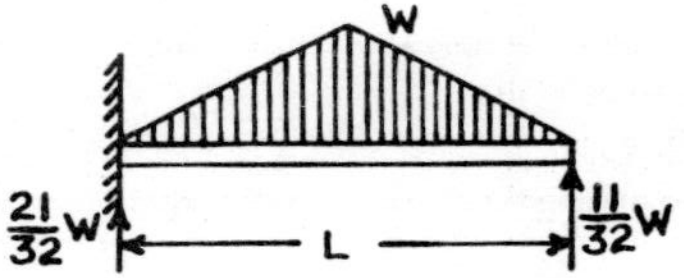	$\frac{5}{32}$WL(+ve)	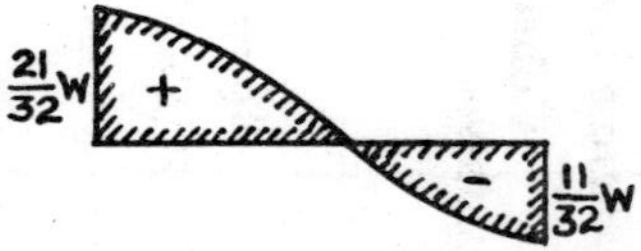	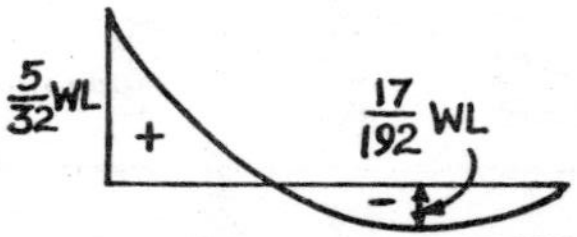	$\frac{1}{139.5}\cdot\frac{WL^3}{EI}$

Work = force × distance traversed ;

= weight × height raised ;

Power = amount of work done uniformly for the unit of time.

$$\text{Mechanical advantage} = \frac{\text{Resistance overcome}}{\text{effort applied}} ;$$

$$\text{Velocity ratio} = \frac{\text{distance moved by effort}}{\text{distance through which the resistance is overcome}} ;$$

$$\text{Efficiency of a machine} = \frac{\text{Mechanical advantage}}{\text{Velocity ratio}} ;$$

First system of Pulleys :

Neglecting weights of pulleys, Mechanical advantage of the system

$$= \frac{W}{P} = 2^n ;$$

where n is the number of pulleys.

Considering weights of pulleys as w_1, w_2, w_3, etc., for n number of pulleys

$$P = \frac{W}{2^n} + \frac{w_1}{2^n} + \frac{w_2}{2^{n-1}} + \ldots\ldots + \frac{w_n}{2}$$

If all the pulleys are equal in weight

$$2^n P = W + w(2^n - 1) ;$$

where w = wt. of each pulley.

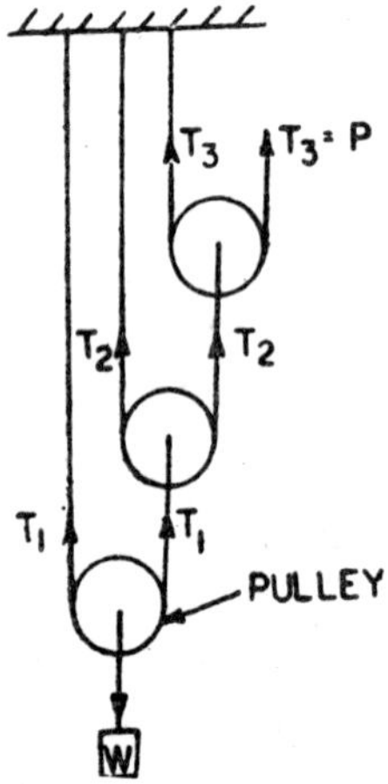

Second System of Pulleys :

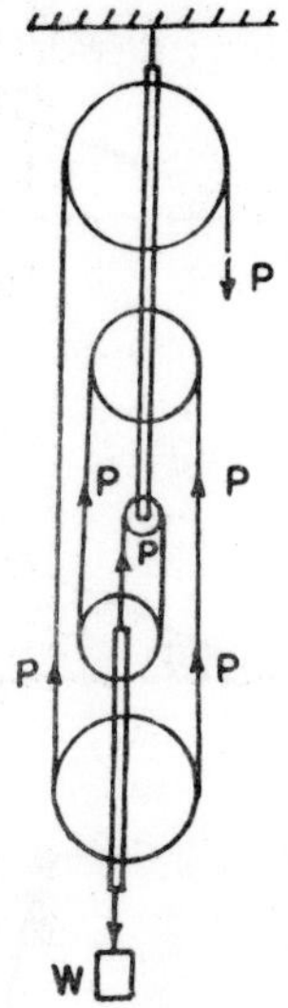

Let n=number of portions of string at the lower block,

W=weight supported, and

w=weight of the lower block :

Then, $W+w=nP$

Third System of Pulleys :

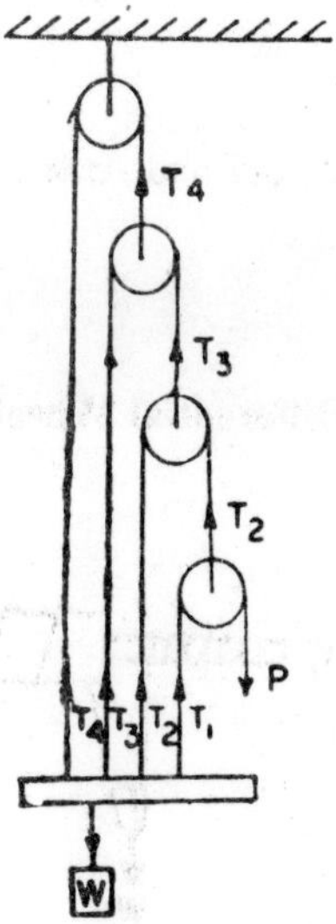

Neglecting weights of pulleys, for n number of pulleys

$$W=P(2^n-1);$$

Considering weights of pulleys,

$$W=(2^n-1)P+w(2^n-n-1);$$

where n=no. of pulleys,

and w=wf. of each pulley.

Inclined Plane :

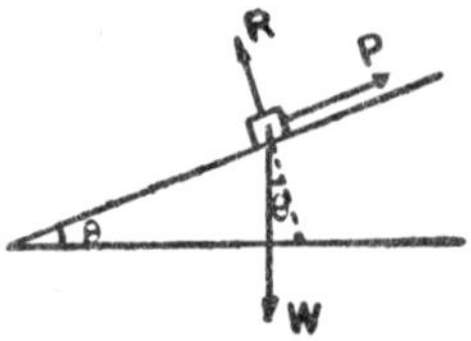

$P = W \sin\theta$;
$R = W \cos\theta$;

$$\frac{P}{\sin\theta} = \frac{R}{\cos\theta} = W ;$$

or, $P : R : W$

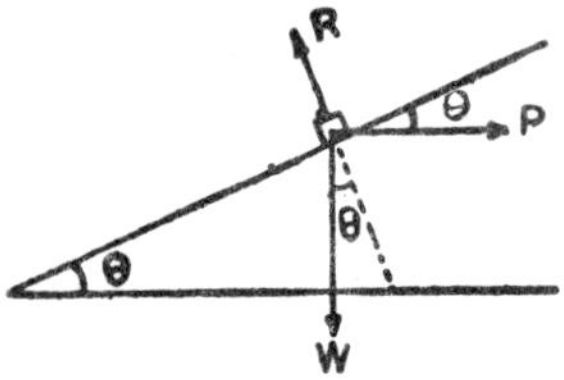

$P = W \tan\theta$:

$R = W \sec\theta$;

$$\frac{P}{\sin\theta} = \frac{R}{1} = \frac{W}{\cos\theta} ;$$

or, $P : R : W$

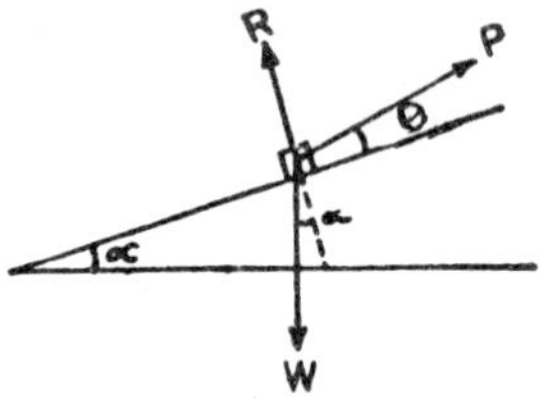

$$P = W \frac{\sin\alpha}{\cos\theta} ;$$

$$\frac{P}{\sin\alpha} = \frac{R}{\cos(\theta+\alpha)} = \frac{W}{\cos\theta}$$

$$R = W \frac{\cos(\theta+\alpha)}{\cos\theta} ;$$

Wheel and Axle : Mechanical advantage

$$= \frac{\text{radius of the wheel}}{\text{radius of the axle}}$$

Differential Wheel and Axle :

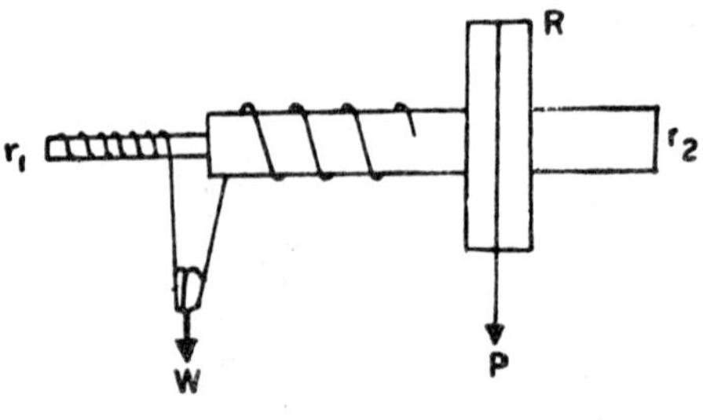

Mechanical advantage

$$= \frac{W}{P} \quad \frac{2R}{r_2 - r_1}$$

Weston's Differential Pulley :

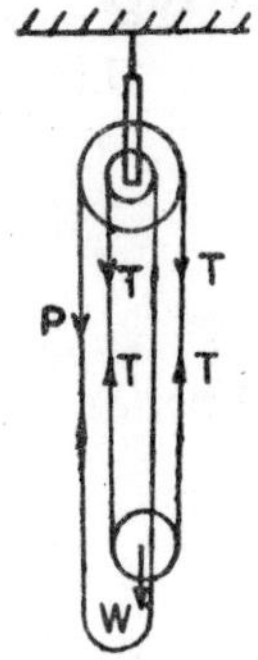

The chain is endless.

Let R and r be the radii of the larger and smaller pulleys of the upper block.

The mechanical advantage of this system

$$=\frac{W}{P}=\frac{2R}{R-r}$$

Screw :

Mechanical advantage

$$=\frac{\text{circumference of a circle whose radius is the effort arm}}{\text{distance between the consecutive threads of the screw}}$$

Hunter's Differential Screw :

Mechanical advantage

$$=\frac{\text{circumference of the circle described by the end of the power arm}}{\text{difference of the pitches of the two screws}}.$$

Friction :

Substance	*Co-efficient of friction*	*Angle of friction*
Wood on wood	·25 to ·5	14° to $26\frac{1}{2}$°
Metals on metals (Dry)	·15 to ·2	$8\frac{1}{2}$° to $11\frac{1}{2}$°
Metals on metals (Wet)	·3	$16\frac{1}{2}$°
Leather on metals (Dry)	·56	$29\frac{1}{2}$°
Leather on metals (Wet)	·36	20°
Leather on metals (Oily)	·15	$8\frac{1}{2}$°

Wedge :

Let α=angle of friction

$$\frac{2R}{P}=\frac{\cos\alpha}{\sin\left(\frac{\theta}{2}+\alpha\right)}$$

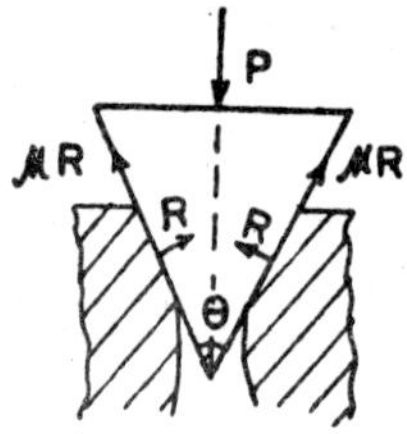

Friction :

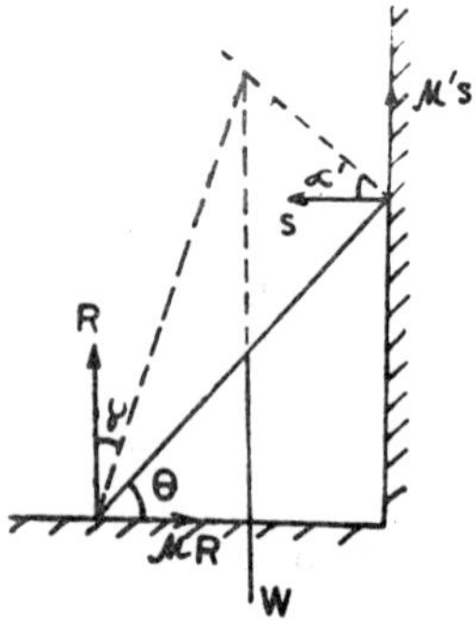

$$\mu R=S$$

$$R+\mu' S=W$$

$$\tan\theta=\frac{1-\mu\mu'}{2\mu}$$

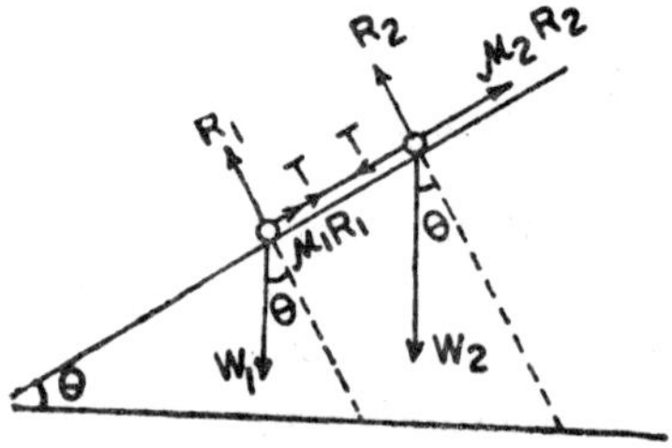

$$\tan\theta=\frac{W_1\mu_1+W_2\mu_2}{W_1+W_2}$$

10

Dynamics

Resultant of two velocities :

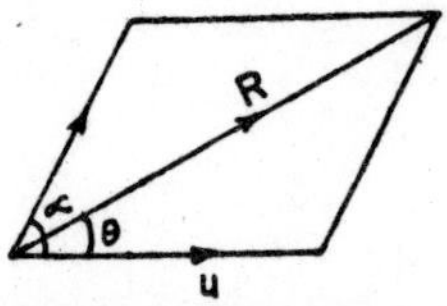

$$R=\sqrt{u^2+v^2+2uv\cos\alpha}$$

$$\theta=\tan^{-1}\frac{v\sin\alpha}{u+v\cos\alpha}$$

If $\alpha=0$, $R=u+v$ and if $\alpha=180°$, $R=u-v$

$$u=\frac{R\sin(\alpha-\theta)}{\sin\alpha} \text{ and } v=\frac{R\sin\theta}{\sin\alpha}$$

Uniformly accelerated motion along a straight path :

(*i*) $v=u+ft$ where. u=initial velocity

(*ii*) $s=ut+\frac{1}{2}ft^2$ v=final velocity

(*iii*) $v^2=u^2+2fs$ f=uniform acceleration

t=time interval between u and v.

and, s=distance traversed in time t.

Distance traversed in any particular time :

$$s_t=u+\tfrac{1}{2}f(2t-1)$$

Body moving vertically downwards :

Time taken to reach the ground, $t=\sqrt{\dfrac{2h}{g}}$

Velocity on reaching the ground, $v=\sqrt{2gh}$

where, h=height through which the body is falling ;

g=acceleration due to gravity.

Body projected vertically upwards :

Time of rise to the maximum height, $T=\dfrac{u}{g}$.

Max. height attained by the body, $H=\dfrac{u^2}{2g}$.

Time of rise=time of fall.

Thus, total time of flight$=\dfrac{2u}{g}$.

Body sliding down a plane :

Velocity on reaching the bottom, $v=\sqrt{2\ gl \sin \alpha}$.

where, l=length traversed by the body.

and, α=inclination of the plane to the horizontal.

Body projected up an inclined plane :

Time of rise to the max. height, $T=\dfrac{u}{g \sin \alpha}$.

Length described by the body, $l=\dfrac{u^2}{2g \sin \alpha}$.

Max. vertical height attained, $H=\dfrac{u^2}{2g}$.

Time of rise=time of fall.

Motion of a body down a chord of a vertical circle :

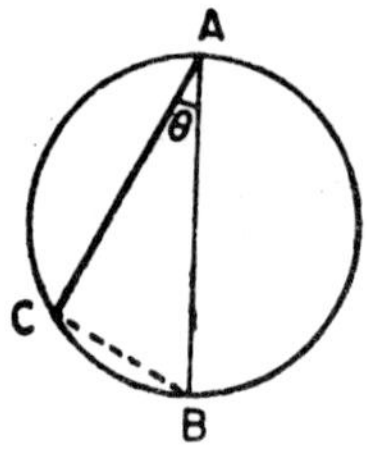

Time of sliding down AC,

$$T=\sqrt{\frac{2d}{g}}$$

where, d=diameter of the circle.

$AC=d \cos \theta$

Projectiles :

Total time of flight,

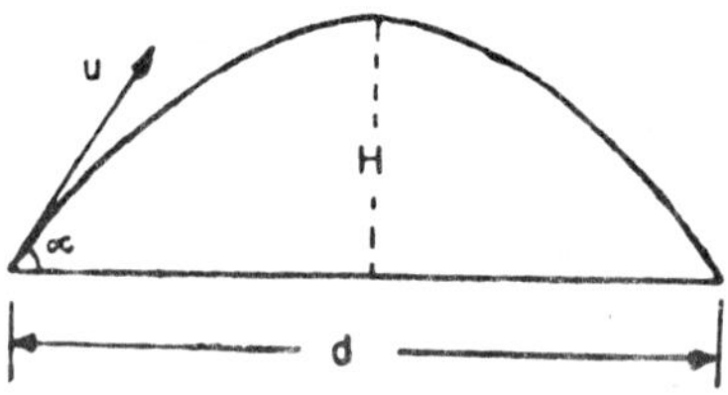

$$T=\frac{2\ u \sin\ \alpha.}{g}$$

$$d=\frac{u^2}{g}\sin 2\alpha\ .$$

Greatest height attained by the projectile, $H=\dfrac{u^2 \sin^2 \alpha}{2g}$

Time taken to attain greatest height 'T'$=\dfrac{u \sin \alpha}{g}$

Motion of connected systems :

$$f=\frac{m_1-m_2}{m_1+m_2}\cdot g$$

$$T=\frac{2m_1m_2}{m_1+m_2}\cdot g$$

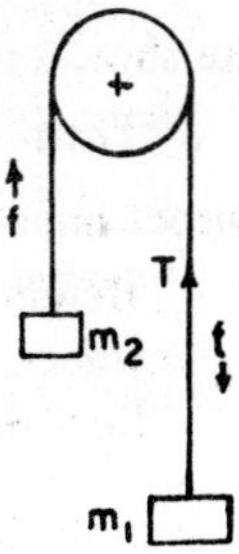

$$f=\frac{m_1}{m_1+m_2}\cdot g$$

$$T=\frac{m_1m_2}{m_1+m_2}\cdot g$$

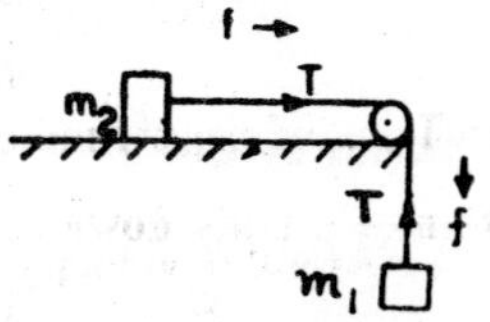

$$f=\frac{m_1-m_2\sin\alpha}{m_1+m_2}\cdot g$$

$$T=\frac{m_1\,m_2\,(1+\sin\alpha)}{m_1+m_2}\cdot g$$

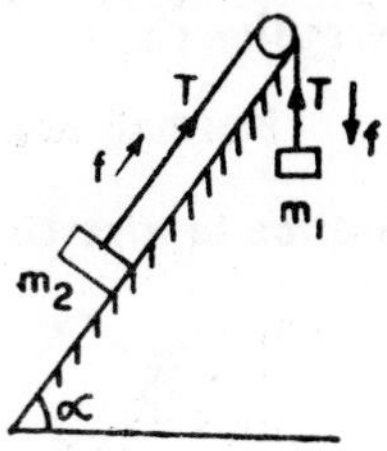

Impulse :

Impulse, $I=Pt=m(v-u)$

where, P=force acting on a body of mass m for any time t,
u=initial velocity,

and, v=final velocity at the end of t.

Collision of two bodies :

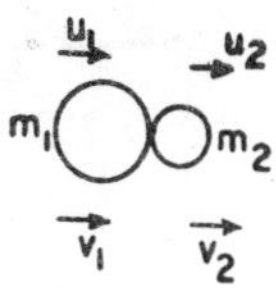

$$m_1v_1+m_2v_2=m_1u_1+m_2u_2$$

co-eff. of restitution, $e=\dfrac{v_2-v_1}{u_1-u_2}$

Motion of a shot and a gun :

Let m=mass of shot, M=mass of gun, v=muzzle velocity of the shot, and, V=recoil velocity of the gun.

Then, $MV=mv$

Direct impact of a sphere on a plane :

Height to which the sphere rebounds$=e^2h$
where, h=height through which the sphere falls.
Loss of kinetic energy due to collision (direct impact)

$$=\frac{1}{2}\frac{m_1m_2}{m_1+m_2}(1-e^2)(u_1-u_2)^2$$

Impulsive action between two colliding spheres is given by

$$I=\frac{m_1m_2}{m_1+m_2}(1+e)(u_1-u_2)$$

Circular motion :

Angular velocity, $w=\frac{v}{r}$;

where, v=uniform speed of a particle moving round a circle of radius r.

Normal acceleration, $\alpha=\frac{v^2}{r}$:

Motion in a vertical circle :

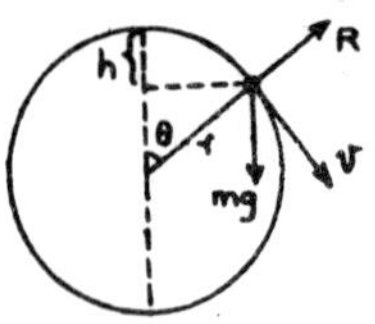

(*i*) When the particle slides down the outside of the arc of the vertical circle from the highest point,

$$R=\frac{mg}{r}(r-3h)$$

$$v=\sqrt{2\,gr\,(1-\cos\theta)}$$

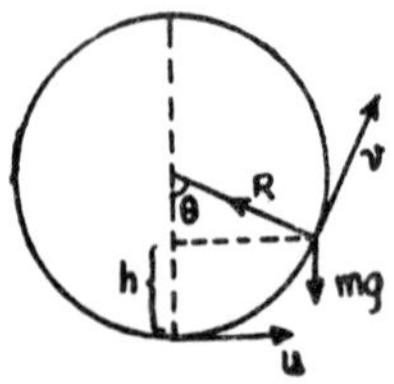

(*ii*) When the particle is projected from the lowest point to move along the inside of the circular arc,

$$R=\frac{m}{r}\{u^2-2gr+3gr\cos\theta\}$$

$$v=\sqrt{u^2-2\,gr+2\,gr\cos\theta}$$

11

Units

Units of Length :

British Units

12 inches	=1 foot
3 feet	=1 yard
220 yards	=1 furlong
8 furlongs	=1 mile

Metric Units

10 millimetres	=1 centimetre
10 centimetres	=1 decimetre
10 decimetres	=1 metre
10 metres	=1 dekametre
10 dekametres	=1 hectometre
10 hectometres	=1 kilometre

1 metre=100 cm.=1000 mm.

1 kilometre =1000 metres

Conversions of British Units to Metric Units

1 inch	=25·4 millimetres
1 foot	=30·48 centimetres
1 yard	=0·9144 metre
1 mile	=1·6093 kilometres

Units of Area :

British Units

144 square inches	=1 square foot
9 square feet	=1 square yard

4840 square yards	=1 acre
640 acres	=1 square mile

Metric Units

100 square millimetres	=1 square centimetre
100 square centimetres	=1 square decimetre
100 square decimetres	=1 square metre
100 square metres	=1 square dekametre
100 square dekametres	=1 square hectometre
100 square hectometres	=1 square kilometre

Conversions of British Units to Metric Units

1 square inch	=6·45 square centimetres
1 square foot	=9·29 square decimetres
1 square yard	=0·84 square metre
1 acre	=0·405 square hectometre
1 square mile	=2·59 square kilometres

Units of Volume :

British Units

1,728 cu. in.	=1 cu. ft.
27 cu. ft.	=1 cu. yd.

Metric Units

1,000 cu. mm.	=1 cu. cm.
1,000 cu. cm.	=1 cu. decimetre
1,000 cu. dm.	=1 cu. m.

Units of Capacity :

British Units

60 minims	=1 fluid drachm
8 fluid drachms	=1 fluid ounce
5 fluid ounces	=1 gill
4 gills	=1 pints

2 pints	=1 quart
4 quarts	=1 gallon

Metric Units

10 millilitres	=1 centilitre
10 centilitres	=1 decilitre
10 decilitres	=1 litre
1,000 litres	=1 kilolitre

Units of Weight :

Metric Units

10 milligrams	=1 centigram
10 centigrams	=1 decigram
10 decigrams	=1 gram
1,000 grams	=1 kilogram
100 kilograms	=1 quintal
1,000 kilograms	=1 tonne

INCHES TO MILLIMETRES

1 to 100 inches

in	*mm*	*in*	*mm*	*in*	*mm*	*in*	*mm*
1	25·4	26	660·4	51	1295·4	76	1930·4
2	50·8	27	685·8	52	1320·8	77	1955·8
3	76·2	28	711·2	53	1346·2	78	1981·2
4	101·6	29	736·6	54	1371·6	79	2006·6
5	127·0	30	762·0	55	1397·0	80	2032·0
6	152·4	31	787·4	56	1422·4	81	2057·4
7	177·8	32	812·8	57	1447·8	82	2082·8
8	203·2	33	838·2	58	1473·2	83	2108·2
9	228·6	34	863·6	59	1498·6	84	2133·6
10	254·0	35	889·0	60	1524·0	85	2159·0
11	297·4	36	914·4	61	1549·4	86	2184·4
12	304·8	37	939·8	62	1574·8	87	2209·8
13	330·2	38	965·2	63	1600·2	88	2235·2
14	355·6	39	990·6	64	1625·6	89	2260·6
15	381·0	40	1016·0	65	1651·0	90	2286·0
16	406·4	41	1041·4	66	1676·4	91	2311·4
17	431·8	42	1066·8	67	1701·8	92	2336·8
18	457·2	43	1092·2	68	1727·2	93	2362·2
19	482·6	44	1117·6	69	1752·6	94	2387·6
20	508·0	45	1143·0	70	1778·0	95	2413·0
21	533·4	46	1168·4	71	1803·4	96	2438·4
22	558·8	47	1193·8	72	1828·8	97	2463·8
23	584·2	48	1219·2	73	1854 2	98	2489·2
24	609·6	49	1244·6	74	1879·6	99	2514·6
25	635·0	50	1270·0	75	1905·0	100	2540·0

MILLIMETRES TO INCHES
1 to 100 Millimetres

mm	*in*	*mm*	*in*	*mm*	*in*	*mm*	*in*
1	0·039370	26	1·023622	51	2·007874	76	2·992126
2	0·078740	27	1·062992	52	2·047244	77	3·031496
3	0·118110	28	1·102362	53	2·086614	78	3·070866
4	0·157480	29	1·141732	54	2·125984	79	3·110236
5	0·196850	30	1·181102	55	2·165354	80	3·149606
6	0·236220	31	1·220472	56	2·204724	81	3·188976
7	0·275591	32	1·259843	57	2·244095	82	3·228346
8	0·314961	33	1·299213	58	2·283465	83	3·267717
9	0·354331	34	1·338583	59	2·322835	84	3·307087
10	0·393701	35	1·377953	60	2·362205	85	3·346457
11	0·433071	36	1·417323	61	2·401575	86	3·385827
12	0·472441	37	1·456693	62	2·440945	87	3·425197
13	0·511811	38	1·496063	63	2·480315	88	3·464567
14	0·551181	39	1·535433	64	2·519685	89	3·503937
15	0·590551	40	1·574803	65	2·559055	90	3·543307
16	0·629921	41	1·614173	66	2·598425	91	3·582677
17	0·669291	42	1·653543	67	2·637795	92	3·622047
18	0·708661	43	1·692913	68	2·677165	93	3·661417
19	0·748032	44	1·732283	69	2·716535	94	3·700787
20	0·787402	45	1·771654	70	2·755906	95	3·740158
21	0·826772	46	1·811024	71	2·795276	96	3·779528
22	0·866142	47	1·850394	72	2·834646	97	3·818898
23	0·905512	48	1·889764	73	2·874016	98	3·858268
24	0·944882	49	1·929134	74	2·913386	99	3·897638
25	0·984252	50	1·968504	75	2·952756	100	3·937008

INCHES WITH FRACTIONS OF AN INCH TO MILLIMETERS

Inches	0	$\frac{1}{16}''$	$\frac{1}{8}''$	$\frac{3}{16}''$	$\frac{1}{4}''$	$\frac{5}{16}''$	$\frac{6}{16}''$	$\frac{7}{16}''$	$\frac{1}{2}''$	$\frac{9}{16}''$	$\frac{5}{8}''$	$\frac{11}{16}''$	$\frac{3}{4}''$	$\frac{13}{16}''$	$\frac{7}{8}''$	$\frac{15}{16}''$
0	0·0	1·6	3·2	4·8	6·3	7·9	9·5	11·1	12·7	14·3	15·9	17·5	19·0	20·6	22·2	23·8
1″	25·4	27·0	28·6	30·2	31·7	33·3	34·9	36·5	38·1	39·7	41·3	42·9	44·4	46·0	47·6	49·2
2′	50·8	52·4	54·0	55·6	57·1	58·7	60·3	61·9	63·5	65·1	66·7	68·3	69·8	71·4	73·0	74·6
3″	76·2	77·8	79·4	81·0	82·5	84·1	85·7	87·3	88·9	90·5	92·1	93·7	95·2	96·8	98·4	100·0
4″	101·6	103·2	104·8	106·4	107·9	109·5	111·1	112·7	114·3	115·9	117·5	119·1	120·6	122·2	123·8	125·4
5″	127·0	128·6	130·2	131·8	133·3	134·9	136·5	138·1	139·7	141·3	142·9	144·5	146·0	147·6	149·2	150·8
6″	152·4	154·0	155·6	157·2	158·7	160·3	161·9	163·5	164·1	166·7	168·3	169·9	171·4	173·0	174·6	176·2
7″	177·8	179·4	181·0	182·6	184·1	185·7	187·3	188·9	190·5	192·1	193·7	195·3	196·8	198·4	200·0	201·6
8″	203·2	204·8	206·4	208·0	209·5	211·1	212·7	214·3	215·9	217·5	219·1	220·7	222·2	223·8	225·4	227·0
9″	228·6	230·2	231·8	233·4	234·9	236·5	238·1	239·7	241·3	224·9	244·5	246·1	247·6	249·2	250·8	252·4
10″	254·0	255·6	[illegible]	[illegible]	[illegible]	[illegible]	[illegible]	[illegible]	[illegible]	[illegible]	[illegible]	[illegible]	[illegible]	[illegible]	[illegible]	[illegible]

EQUIVALENT METRIC UNITS AND CONVERSION FACTORS

Quantity	British Unit	Equivalent Metric Unit	Conversion Factor: British to Metric	Conversion Factor: Metric to British
Length	inch (in)	millimetre (mm)	25.400	0.039
	inch [in)	centimetre (cm)	2.540	0.394
	foot (ft)	metre (m)	0.305	3.281
	yard (yd)	metre (m)	0.914	1.094
	furlong	kilometre (km)	0.201	4.971
	mile	kilometre (km)	1.609	0.621
	natutical mile (U.K.)	international nautical mile (n mile)	1.001	0.999
Area	square inch (in^2)	square centimetre (cm^2)	6.452	0.155
	square foot (ft^2)	square metre (m^2)	0.0929	10.764
	square yard (yd^2)	square metre (m^2)	0.836	1.196
	acre	hectare [ha]	0.405	2.471
	square mile (sq mile)	square kilometre (km^2)	2.590	0.386
Volume	cubic inch (in^3)	cubic centimetre (cm^3)	16.387	0.061
	cubic foot (ft^3)	cubic metre (m^3)	0.0283	35.315
	cubic yard $(yd)^3$	cubic metre (m^3)	0.765	1.308
	acre foot (acre ft)	hectare metre (ha m)	0.123	8.107
Capacity	imperial gallon (gal)	litre (1)	4.546	0.220

Mass	pound (lb) hundred-weight ton	kilogram (kg) quintal (q) metric tonne (t)	0.454 0.508 1.016	2.205 1.968 0.984
Linear Density	lb/ft	kg/m	1.488	0.672
Density (Mass/Volume)	lb/ft^3 lb/in^3	kg/m^3 g/cm^3	16.019 27.680	0.062 0.036
Specific Volume	ft^3/lb	m^3/kg	0.062 4	16.019
Momentum	lb ft/s	kg m/s	0.138	7.233
Moment of Momentum	lb ft^2/s	kg m^2/s	0.042 1	23.730
Moment of Inertia	in^4	cm^4	41.623	0.024
Section Modulus	in^3	cm^3	16.387	0.061
Force	lbf	kgf	0.454	2.205
Bending Moment	lbf ft lbf in	kgf m kgf cm	0.138 1.152	7.233 0.868

Stress and Pressure	lbf/in² lbf/ft² tonf/ft² tonf/in²	kgf/cm² kgf/m² tf/m² kgf/mm²	0.070 3 4.882 10.937 1.575	14.223 0.205 0.091 0.635
Dynamic Viscosity	Poundal second square foot (pdl s/ft²) = lb/ft/s	Poise (P) = dyn s/cm² = g/cm s	14.882	0.067
Kinematic Viscosity	in²/s ft²/s	Stokes (S) = cm²/s Stokes (S) = cm²/s	6.452 929.034	0.155 0.001
Work and Energy	ft lbf	kgf m	0.138	7.233
Power	horse power (hp) horse power (hp)	kgf m/s kilowatt (kW)	76.040 0.746	0.013 2 1.341
Temperature	⁰F	⁰C	$(F-32)\ \frac{5}{9}$	$\left\{C\times \frac{9}{5}+32\right\}$
Quantity of Heat	British thermal unit (Btu)	Calorie (cal)	251.996	0.004
Heat Flow Rate	Btu/h Btu/h	cal/s kcal/h	0.070 0.252	14.286 3.968
Density of Heat Flow Rate	Btu/ft²h	kcal/m²h	2.712	0.369

Thermal Conductivity	Btu/h ft deg F Btu in/h ft² deg F	kcal/h m deg C kcal/h m deg C	1.488 0.124	0.672 8.064
Thermal Coductance	Btu/ft² h deg F	kcal/m²h deg C	4.882	0.205
Calorific Value	Btu/lb Btu/lb	J/g cal/g	2.326 0.556	0.430 1.800
Specific Heat	Btu/lb deg F Btu/lb deg F	J/g deg C cal/g deg C	4.187 1.000	0.239 1.000
Consumption (Volume/Length)	gallon per mile [gal/mile]	litre per kilometre (l/km)	2.825	0.354
Consumption (Length/Volume)	mile per gallon (mile/gal)	kilometre per litre (km/l)	0.354	2.825
Illumination	foot-candle	lux (lx)	10.764	0.093
Brightness	candela per sq in (cd/in²) foot-lambert	stilb (sb) lambert (L)	0.155 0.001 076	6.452 929.030

Barlow's Tables

SQUARES

CUBES

SQUARE ROOTS

CUBE ROOTS

AND

RECIPROCALS

OF ALL INTEGERS UP TO

1,000

No. n	Square n^2	Cube n^3	Fourth power n^4	$\frac{1}{\sqrt{n}}$	Factorial* $n!$	Sq. root $\sqrt{n}$	Cube root $\sqrt[3]{n}$	Reciprocal $\frac{1}{n}$
				0·				0·
1	1	1	1		1	1·0000000	1·0000000	
2	4	8	16	7071068	2	1·4142136	1·2599210	500000000
3	9	27	81	5773503	6	1·7320508	1·4422496	333333333
4	16	64	256	5000000	24	2·0000000	1·5874011	250000000
5	25	125	625	4472136	120	2·2360680	1·7099759	200000000
6	36	216	1296	4082483	720	2·4494897	1·8171206	166666667
7	49	343	2401	3779645	5040	2·6457513	1·9129312	142857143
8	64	512	4096	3535534	40320	2·8284271	2·0000000	125000000
9	81	729	6561	3333333	362880	3·0000000	2·0800838	111111111
10	1 00	1 000	1 0000	3162278	0·3628800 7	3·1622777	2·1544347	100000000
11	1 21	1 331	1 4641	3015113	3·9916800 7	3·3166248	2·2239801	090909091
12	1 44	1 728	2 0736	2886751	4·7900160 8	3·4641016	2·2894285	083333333
13	1 69	2 197	2 8561	2773501	6·2270208 9	3·6055513	2·3513347	076923077
14	1 96	2 744	3 8416	2672612	8·7178291 10	3·7416574	2·4101423	071428571
15	2 25	3 375	5 0625	2581989	1·3076744 12	3·8729833	2·4662121	066666667
16	2 56	4 096	6 5536	2500000	2·0922790 13	4·0000000	2·5198421	062500000
17	2 89	4 913	8 3521	2425356	3·5568743 14	4·1231056	2·5712816	058823529
18	3 24	5 832	10 4976	2357023	6·4023737 15	4·2426407	2·6207414	055555556
19	3 61	6 859	13 0321	2294157	1·2164510 17	4·3588989	2·6684016	052631579
20	4 00	8 000	16 0000	2236068	2·4329020 18	4·4721360	2·7144176	050000000
21	4 41	9 261	19 4481	2182179	5·1090942 19	4·5825757	2·7589242	047619048
22	4 84	10 648	23 4256	2132007	1·1240007 21	4·6904158	2·8020393	045454545
23	5 29	12 167	27 9841	2085144	2·5852017 22	4·7958315	2·8438670	043478261
24	5 76	13 824	33 1776	2041241	6·2044840 23	4·8989795	2·8844991	041666667
25	6 25	15 625	39 0625	2000000	1·5511210 25	5·0000000	2·9240177	040000000
26	6 76	17 576	45 6976	1961161	4·0329146 26	5·0990195	2·9624961	038461538
27	7 29	19 683	53 1441	1924501	1·0888869 28	5·1961524	3·0000000	037037037
28	7 84	21 952	61 4656	1889822	3·0488834 29	5·2915026	3·0365890	035714286
29	8 41	24 389	70 7281	1856953	8·8417620 30	5·3851648	3·0723168	034482759
30	9 00	27 000	81 0000	1825742	2·6525286 32	5·4772256	3·1072325	033333333
31	9 61	29 791	92 3521	1796053	8·2228387 33	5·5677644	3·1413807	032258065
32	10 24	32 768	104 8576	1767767	2·6313084 35	5·6568542	3·1748021	031250000
33	10 89	35 937	118 5921	1740777	8·6833176 36	5·7445626	3·2075343	030303030
34	11 56	39 304	133 6336	1714986	2·9523280 38	5·8309519	3·2396118	029411765
35	12 25	42 875	150 0625	1690309	1·0333148 40	5·9160798	3·2710663	028571429
36	12 96	46 656	167 9616	1666667	3·7199333 41	6·0000000	3·3019272	027777778
37	13 69	50 653	187 4161	1643990	1·3763753 43	6·0827625	3·3322219	027027027
38	14 44	54 872	208 5136	1622214	5·2302262 44	6·1644140	3·3619754	026315789
39	15 21	59 319	231 3441	1601282	2·0397882 46	6·2449980	3·3912114	025641026
40	16 00	64 000	256 0000	1581139	8·1591528 47	6·3245553	3·4199519	025000000
41	16 81	68 921	282 5761	1561738	3·3452527 49	6·4031242	3·4482172	024390244
42	17 64	74 088	311 1696	1543034	1·4050061 51	6·4807407	3·4760266	023809524
43	18 49	79 507	341 8801	1524986	6·0415263 52	6·5574385	3·5033981	023255814
44	19 36	85 184	374 8096	1507557	2·6582716 54	6·6332496	3·5303483	022727273
45	20 25	91 125	410 0625	1490712	1·1962222 56	6·7082039	3·5568933	022222222
46	21 16	97 336	447 7456	1474420	5·5026222 57	6·7823300	3·5830479	021739130
47	22 09	103 823	487 9681	1458650	2·5862324 59	6·8556546	3·6088261	021276596
48	23 04	110 592	530 8416	1443376	1·2413916 61	6·9282032	3·6342412	020833333
49	24 01	117 649	576 4801	1428571	6·0828186 62	7·0000000	3·6593057	020408163
50	25 00	125 000	625 0000	1414214	3·0414093 64	7·0710678	3·6840315	020000000

*The number following $n!$ is the power of 10 by which the given tabular quantity must be multiplied.

No. n	Square n^2	Cube n^3	Fourth power n^4	$\frac{1}{\sqrt{n}}$	Factorial* $n!$	Sq. root $\sqrt{n}$	Cube root $\sqrt[3]{n}$	Reciprocal $\frac{1}{n}$
				0·				0·0
50	25 00	125 000	625 0000	1414214	3·0414093 64	7·0710678	3·6840315	20000000
51	26 01	132 651	676 5201	1400280	1·5511188 66	7·1414284	3·7084298	19607843
52	27 04	140 608	731 1616	1386750	8·0658175 67	7·2111026	3·7325112	19230769
53	28 09	148 877	789 0481	1373606	4·2748833 69	7·2801099	3·7562858	18867925
54	29 16	157 464	850 3056	1360828	2·3084370 71	7·3484692	3·7797631	18518519
55	30 25	166 375	915 0625	1348400	1·2696403 73	7·4161985	3·8029525	18181818
56	31 36	175 616	983 4496	1336306	7·1099859 74	7·4833148	3·8258624	17857143
57	32 49	185 193	1055 6001	1324532	4·0526920 76	7·5498344	3·8485011	17543860
58	33 64	195 112	1131 6496	1313064	2·3505613 78	7·6157731	3·8708766	17241379
59	34 81	205 379	1211 7361	1301889	1·3868312 80	7·6811457	3·8929964	16949153
60	36 00	216 000	1296 0000	1290994	8·3209871 81	7·7459667	3·9148676	16666667
61	37 21	226 981	1384 5841	1280369	5·0758021 83	7·8102497	3·9364972	16393443
62	38 44	238 328	1477 6336	1270001	3·1469973 85	7·8740079	3·9578916	16129032
63	39 69	250 047	1575 2961	1259882	1·9826083 87	7·9372539	3·9790572	15873016
64	40 96	262 144	1677 7216	1250000	1·2688693 89	8·0000000	4·0000000	15625000
65	42 25	274 625	1785 0625	1240347	8·2476506 90	8·0622577	4·0207258	15384615
66	43 56	287 496	1897 4736	1230915	5·4434494 92	8·1240384	4·0412400	15151515
67	44 89	300 763	2015 1121	1221694	3·6471111 94	8·1853528	4·0615481	14925373
68	46 24	314 432	2138 1376	1212678	2·4800355 96	8·2462113	4·0816551	14705882
69	47 61	328 509	2266 7121	1203859	1·7112245 98	8·3066239	4·1015659	14492754
70	49 00	343 000	2401 0000	1195229	1·1978572 100	8·3666003	4·1212853	14285714
71	50 41	357 911	2541 1681	1186782	8·5047859 101	8·4261498	4·1408177	14084507
72	51 84	373 248	2687 3856	1178511	6·1234458 103	8·4852814	4·1601676	13888889
73	53 29	389 017	2839 8241	1170411	4·4701155 105	8·5440037	4·1793392	13698630
74	54 76	405 224	2998 6576	1162476	3·3078854 107	8·6023253	4·1983365	13513514
75	56 25	421 875	3164 0625	1154701	2·4809141 109	8·6602540	4·2171633	13333333
76	57 76	438 976	3336 2176	1147079	1·8854947 111	8·7177979	4·2358236	13157895
77	59 29	456 533	3515 3041	1139606	1·4518309 113	8·7749644	4·2543209	12987013
78	60 84	474 552	3701 5056	1132277	1·1324281 115	8·8317609	4·2726587	12820513
79	62 41	493 039	3895 0081	1125088	8·9461821 116	8·8881944	4·2908404	12658228
80	64 00	512 000	4096 0000	1118034	7·1569457 118	8·9442719	4·3088694	12500000
81	65 61	531 441	4304 6721	1111111	5·7971260 120	9·0000000	4·3267487	12345679
82	67 24	551 368	4521 2176	1104315	4·7536433 122	9·0553851	4·3444815	12195122
83	68 89	571 787	4745 8321	1097643	3·9455240 124	9·1104336	4·3620707	12048193
84	70 56	592 704	4978 7136	1091089	3·3142401 126	9·1651514	4·3795191	11904762
85	72 25	614 125	5220 0625	1084652	2·8171041 128	9·2195445	4·3968297	11764706
86	73 96	636 056	5470 0816	1078328	2·4227095 130	9·2736185	4·4140050	11627907
87	75 69	658 503	5728 9761	1072113	2·1077573 132	9·3273791	4·4310476	11494253
88	77 44	681 472	5996 9536	1066004	1·8548264 134	9·3808315	4·4479602	11363636
89	79 21	704 969	6274 2241	1059998	1·6507955 136	9·4339811	4·4647451	11235955
90	81 00	729 000	6561 0000	1054093	1·4857160 138	9·4868330	4·4814047	11111111
91	82 81	753 571	6857 4961	1048285	1·3520015 140	9·5393920	4·4979414	10989011
92	84 64	778 688	7163 9296	1042572	1·2438414 142	9·5916630	4·5143574	10869565
93	86 49	804 357	7480 5201	1036952	1·1567725 144	9·6436508	4·5306549	10752688
94	88 36	830 584	7807 4896	1031421	1·0873662 146	9·6953597	4·5468359	10638298
95	90 25	857 375	8145 0625	1025978	1·0329978 148	9·7467943	4·5629026	10526316
96	92 16	884 736	8493 4656	1020621	9·9167793 149	9·7979590	4·5788570	10416667
97	94 09	912 673	8852 9281	1015346	9·6192760 151	9·8488578	4·5947009	10309278
98	96 04	941 192	9223 6816	1010153	9·4268904 153	9·8994949	4·6104363	10204082
99	98 01	970 299	9605 9601	1005038	9·3326215 155	9·9498744	4·6260650	10101010
100	100 00	1000 000	10000 0000	1000000	9·3326215 157	10·0000000	4·6415888	10000000

*The number following $n!$ is the power of 10 by which the given tabular quantity must be multiplied.

No. n	Square n^2	Cube n^3	Fourth power n^4	$\frac{1}{\sqrt{n}}$	Square root $\sqrt{n}$	Cube root $\sqrt[3]{n}$	Reciprocal $\frac{1}{n}$
				0·0			0·00
100	1 00 00	1 000 000	1 0000 0000		10·0000000	4·6415888	
101	1 02 01	1 030 301	1 0406 0401	995037	10·0498756	4·6570095	9900990
102	1 04 04	1 061 208	1 0824 3216	990148	10·0995049	4·6723287	9803922
103	1 06 09	1 092 727	1 1255 0881	985329	10·1488916	4·6875481	9708738
104	1 08 16	1 124 864	1 1698 5856	980581	10·1980390	4·7026694	9615385
105	1 10 25	1 157 625	1 2155 0625	975900	10·2469508	4·7176940	9523810
106	1 12 36	1 191 016	1 2624 7696	971286	10·2956301	4·7326235	9433962
107	1 14 49	1 225 043	1 3107 9601	966736	10·3440804	4·7474594	9345794
108	1 16 64	1 259 712	1 3604 8896	962250	10·3923048	4·7622032	9259259
109	1 18 81	1 295 029	1 4115 8161	957826	10·4403065	4·7768562	9174312
110	1 21 00	1 331 000	1 4641 0000	953463	10·4880885	4·7914199	9090909
111	1 23 21	1 367 631	1 5180 7041	949158	10·5356538	4·8058955	9009009
112	1 25 44	1 404 928	1 5735 1936	944911	10·5830052	4·8202845	8928571
113	1 27 69	1 442 897	1 6304 7361	940721	10·6301458	4·8345881	8849558
114	1 29 96	1 481 544	1 6889 6016	936586	10·6770783	4·8488076	8771930
115	1 32 25	1 520 875	1 7490 0625	932505	10·7238053	4·8629441	8695652
116	1 34 56	1 560 896	1 8106 3936	928477	10·7703296	4·8769990	8620690
117	1 36 89	1 601 613	1 8738 8721	924500	10·8166538	4·8909732	8547009
118	1 39 24	1 643 032	1 9387 7776	920575	10·8627805	4·9048681	8474576
119	1 41 61	1 685 159	2 0053 3921	916698	10·9087121	4·9186847	8403361
120	1 44 00	1 728 000	2 0736 0000	912871	10·9544512	4·9324241	8333333
121	1 46 41	1 771 561	2 1435 8881	909091	11·0000000	4·9460874	8264463
122	1 48 84	1 815 848	2 2153 3456	905357	11·0453610	4·9596757	8196721
123	1 51 29	1 860 867	2 2888 6641	901670	11·0905365	4·9731898	8130081
124	1 53 76	1 906 624	2 3642 1376	898027	11·1355287	4·9866310	8064516
125	1 56 25	1 953 125	2 4414 0625	894427	11·1803399	5·0000000	8000000
126	1 58 76	2 000 376	2 5204 7376	890871	11·2249722	5·0132979	7936508
127	1 61 29	2 048 383	2 6014 4641	887357	11·2694277	5·0265257	7874016
128	1 63 84	2 097 152	2 6843 5456	883883	11·3137085	5·0396842	7812500
129	1 66 41	2 146 689	2 7692 2881	880451	11·3578167	5·0527743	7751938
130	1 69 00	2 197 000	2 8561 0000	877058	11·4017543	5·0657970	7692308
131	1 71 61	2 248 091	2 9449 9921	873704	11·4455231	5·0787531	7633588
132	1 74 24	2 299 968	3 0359 5776	870388	11·4891253	5·0916434	7575758
133	1 76 89	2 352 637	3 1290 0721	867110	11·5325626	5·1044687	7518797
134	1 79 56	2 406 104	3 2241 7936	863868	11·5758369	5·1172299	7462687
135	1 82 25	2 460 375	3 3215 0625	860663	11·6189500	5·1299278	7407407
136	1 84 96	2 515 456	3 4210 2016	857493	11·6619038	5·1425632	7352941
137	1 87 69	2 571 353	3 5227 5361	854358	11·7046999	5·1551367	7299270
138	1 90 44	2 628 072	3 6267 3936	851257	11·7473401	5·1676493	7246377
139	1 93 21	2 685 619	3 7330 1041	848189	11·7898261	5·1801015	7194245
140	1 96 00	2 744 000	3 8416 0000	845154	11·8321596	5·1924941	7142857
141	1 98 81	2 803 221	3 9525 4161	842152	11·8743421	5·2048279	7092199
142	2 01 64	2 863 288	4 0658 6896	839181	11·9163753	5·2171034	7042254
143	2 04 49	2 924 207	4 1816 1601	836242	11·9582607	5·2293215	6993007
144	2 07 36	2 985 984	4 2998 1696	833333	12·0000000	5·2414828	6944444
145	2 10 25	3 048 625	4 4205 0625	830455	12·0415946	5·2535879	6896552
146	2 13 16	3 112 136	4 5437 1856	827606	12·0830460	5·2656374	6849315
147	2 16 09	3 176 523	4 6694 8881	824786	12·1243557	5·2776321	6802721
148	2 19 04	3 241 792	4 7978 5216	821995	12·1655251	5·2895725	6756757
149	2 22 01	3 307 949	4 9288 4401	819232	12·2065556	5·3014592	6711409
150	2 25 00	3 375 000	5 0625 0000	816497	12·2474487	5·3132928	6666667

No. n	Square n^2	Cube n^3	Fourth power n^4	$\frac{1}{\sqrt{n}}$	Square root $\sqrt{n}$	Cube root $\sqrt[3]{n}$	Reciprocal $\frac{1}{n}$
				0·0			0·00
150	2 25 00	3 375 000	5 0625 0000	816497	12·2474487	5·3132928	6666667
151	2 28 01	3 442 951	5 1988 5601	813788	12·2882057	5·3250740	6622517
152	2 31 04	3 511 808	5 3379 4816	811107	12·3288280	5·3368033	6578947
153	2 34 09	3 581 577	5 4798 1281	808452	12·3693169	5·3484812	6535948
154	2 37 16	3 652 264	5 6244 8656	805823	12·4096736	5·3601084	6493506
155	2 40 25	3 723 875	5 7720 0625	803219	12·4498996	5·3716854	6451613
156	2 43 36	3 796 416	5 9224 0896	800641	12·4899960	5·3832126	6410256
157	2 46 49	3 869 893	6 0757 3201	798087	12·5299641	5·3946907	6369427
158	2 49 64	3 944 312	6 2320 1296	795557	12·5698051	5·4061202	6329114
159	2 52 81	4 019 679	6 3912 8961	793052	12·6095202	5·4175015	6289308
160	2 56 00	4 096 000	6 5536 0000	790569	12·6491106	5·4288352	6250000
161	2 59 21	4 173 281	6 7189 8241	788110	12·6885775	5·4401218	6211180
162	2 62 44	4 251 528	6 8874 7536	785674	12·7279221	5·4513618	6172840
163	2 65 69	4 330 747	7 0591 1761	783260	12·7671453	5·4625556	6134969
164	2 68 96	4 410 944	7 2339 4816	780869	12·8062485	5·4737037	6097561
165	2 72 25	4 492 125	7 4120 0625	778499	12·8452326	5·4848066	6060606
166	2 75 56	4 574 296	7 5933 3136	776151	12·8840987	5·4958647	6024096
167	2 78 89	4 657 463	7 7779 6321	773823	12·9228480	5·5068784	5988024
168	2 82 24	4 741 632	7 9659 4176	771517	12·9614814	5·5178484	5952381
169	2 85 61	4 826 809	8 1573 0721	769231	13·0000000	5·5287748	5917160
170	2 89 00	4 913 000	8 3521 0000	766965	13·0384048	5·5396583	5882353
171	2 92 41	5 000 211	8 5503 6081	764719	13·0766968	5·5504991	5847953
172	2 95 84	5 088 448	8 7521 3056	762493	13·1148770	5·5612978	5813953
173	2 99 29	5 177 717	8 9574 5041	760286	13·1529464	5·5720547	5780347
174	3 02 76	5 268 024	9 1663 6176	758098	13·1909060	5·5827702	5747126
175	3 06 25	5 359 375	9 3789 0625	755929	13·2287566	5·5934447	5714286
176	3 09 76	5 451 776	9 5951 2576	753778	13·2664992	5·6040787	5681818
177	3 13 29	5 545 233	9 8150 6241	751646	13·3041347	5·6146724	5649718
178	3 16 84	5 639 752	10 0387 5856	749532	13·3416641	5·6252263	5617978
179	3 20 41	5 735 339	10 2662 5681	747435	13·3790882	5·6357408	5586592
180	3 24 00	5 832 000	10 4976 0000	745356	13·4164079	5·6462162	5555556
181	3 27 61	5 929 741	10 7328 3121	743294	13·4536240	5·6566528	5524862
182	3 31 24	6 028 568	10 9719 9376	741249	13·4907376	5·6670511	5494505
183	3 34 89	6 128 487	11 2151 3121	739221	13·5277493	5·6774114	5464481
184	3 38 56	6 229 504	11 4622 8736	737210	13·5646600	5·6877340	5434783
185	3 42 25	6 331 625	11 7135 0625	735215	13·6014705	5·6980192	5405405
186	3 45 96	6 434 856	11 9688 3216	733236	13·6381817	5·7082675	5376344
187	3 49 69	6 539 203	12 2283 0961	731272	13·6747943	5·7184791	5347594
188	3 53 44	6 644 672	12 4919 8336	729325	13·7113092	5·7286543	5319149
189	3 57 21	6 751 269	12 7598 9841	727393	13·7477271	5·7387935	5291005
190	3 61 00	6 859 000	13 0321 0000	725476	13·7840488	5·7488971	5263158
191	3 64 81	6 967 871	13 3086 3361	723575	13·8202750	5·7589652	5235602
192	3 68 64	7 077 888	13 5895 4496	721688	13·8564065	5·7689983	5208333
193	3 72 49	7 189 057	13 8748 8001	719816	13·8924440	5·7789966	5181347
194	3 76 36	7 301 384	14 1646 8496	717958	13·9283883	5·7889604	5154639
195	3 80 25	7 414 875	14 4590 0625	716115	13·9642400	5·7988900	5128205
196	3 84 16	7 529 536	14 7578 9056	714286	14·0000000	5·8087857	5102041
197	3 88 09	7 645 373	15 0613 8481	712470	14·0356688	5·8186479	5076142
198	3 92 04	7 762 392	15 3695 3616	710669	14·0712473	5·8284767	5050505
199	3 96 01	7 880 599	15 6823 9201	708881	14·1067360	5·8382725	5025126
200	4 00 00	8 000 000	16 0000 0000	707107	14·1421356	5·8480355	5000000

No. n	Square n^2	Cube n^3	Fourth power n^4	$\frac{1}{\sqrt{n}}$	Square root $\sqrt{n}$	Cube root $\sqrt[3]{n}$	Reciprocal $\frac{1}{n}$
				0·0			0·00
200	4 00 00	8 000 000	16 0000 0000	707107	14·1421356	5·8480355	5000000
201	4 04 01	8 120 601	16 3224 0801	705346	14·1774469	5·8577660	4975124
202	4 08 04	8 242 408	16 6496 6416	703598	14·2126704	5·8674643	4950495
203	4 12 09	8 365 427	16 9818 1681	701862	14·2478068	5·8771307	4926108
204	4 16 16	8 489 664	17 3189 1456	700140	14·2828569	5·8867653	4901961
205	4 20 25	8 615 125	17 6610 0625	698430	14·3178211	5·8963685	4878049
206	4 24 36	8 741 816	18 0081 4096	696733	14·3527001	5·9059406	4854369
207	4 28 49	8 869 743	18 3603 6801	695048	14·3874946	5·9154817	4830918
208	4 32 64	8 998 912	18 7177 3696	693375	14·4222051	5·9249921	4807692
209	4 36 81	9 129 329	19 0802 9761	691714	14·4568323	5·9344721	4784689
210	4 41 00	9 261 000	19 4481 0000	690066	14·4913767	5·9439220	4761905
211	4 45 21	9 393 931	19 8211 9441	688428	14·[illegible]8390	5·9533418	4739336
212	4 49 44	9 528 128	20 1996 3136	686803	14·5602198	5·9627320	4716981
213	4 53 69	9 663 597	20 5834 6161	685189	14·5945195	5·9720926	4694836
214	4 57 96	9 800 344	20 9727 3616	683586	14·6287388	5·9814240	4672897
215	4 62 25	9 938 375	21 3675 0625	681994	14·6628783	5·9907264	4651163
216	4 66 56	10 077 696	21 7678 2336	680414	14·6969385	6·0000000	4629630
217	4 70 89	10 218 313	22 1737 3921	678844	14·7309199	6·0092450	4608295
218	4 75 24	10 360 232	22 5853 0576	677285	14·7648231	6·0184617	4587156
219	4 79 61	10 503 459	23 0025 7521	675737	14·7986486	6·0276502	4566210
220	4 84 00	10 648 000	23 4256 0000	674200	14·8323970	6·0368107	4545455
221	4 88 41	10 793 861	23 8544 3281	672673	14·8660687	6·0459436	4524887
222	4 92 84	10 941 048	24 2891 2656	671156	14·8996644	6·0550489	4504505
223	4 97 29	11 089 567	24 7297 3441	669650	14·9331845	6·0641270	4484305
224	5 01 76	11 239 424	25 1763 0976	668153	14·9666295	6·0731779	4464286
225	5 06 25	11 390 625	25 6289 0625	666667	15·0000000	6·0822020	4444444
226	5 10 76	11 543 176	26 0875 7776	665190	15·0332964	6·0911993	4424779
227	5 15 29	11 697 083	26 5523 7841	663723	15·0665192	6·1001702	4405286
228	5 19 84	11 852 352	27 0233 6256	662266	15·0996689	6·1091147	4385965
229	5 24 41	12 008 989	27 5005 8481	660819	15·1327460	6·1180332	4366812
230	5 29 00	12 167 000	27 9841 0000	659380	15·1657509	6·1269257	4347826
231	5 33 61	12 326 391	28 4739 6321	657952	15·1986842	6·1357924	4329004
232	5 38 24	12 487 168	28 9702 2976	656532	15·2315462	6·1446337	4310345
233	5 42 89	12 649 337	29 4729 5521	655122	15·2643375	6·1534495	4291845
234	5 47 56	12 812 904	29 9821 9536	653720	15·2970585	6·1622401	4273504
235	5 52 25	12 977 875	30 4980 0625	652328	15·3297097	6·1710058	4255319
236	5 56 96	13 144 256	31 0204 4416	650945	15·3622915	6·1797466	4237288
237	5 61 69	13 312 053	31 5495 6561	649570	15·3948043	6·1884628	4219409
238	5 66 44	13 481 272	32 0854 2736	648204	15·4272486	6·1971544	4201681
239	5 71 21	13 651 919	32 6280 8641	646846	15·4596248	6·2058218	4184100
240	5 76 00	13 824 000	33 1776 0000	645497	15·4919334	6·2144650	4166667
241	5 80 81	13 997 521	33 7340 2561	644157	15·5241747	6·2230843	4149378
242	5 85 64	14 172 488	34 2974 2096	642824	15·5563492	6·2316797	4132231
243	5 90 49	14 348 907	34 8678 4401	641500	15·5884573	6·2402515	4115226
244	5 95 36	14 526 784	35 4453 5296	640184	15·6204994	6·2487998	4098361
245	6 00 25	14 706 125	36 0300 0625	638877	15·6524758	6·2573247	4081633
246	6 05 16	14 886 936	36 6218 6256	637577	15·6843871	6·2658266	4065041
247	6 10 09	15 069 223	37 2209 8081	636285	15·7162336	6·2743054	4048583
248	6 15 04	15 252 992	37 8274 2016	635001	15·7480157	6·2827613	4032258
249	6 20 01	15 438 249	38 4412 4001	633724	15·7797338	6·2911946	4016064
250	6 25 00	15 625 000	39 0625 0000	632456	15·8113883	6·2996052	4000000

No. n	Square n^2	Cube n^3	Fourth power n^4	$\frac{1}{\sqrt{n}}$	Square root $\sqrt{n}$	Cube root $\sqrt[3]{n}$	Reciprocal $\frac{1}{n}$
				0·0			0·00
250	6 25 00	15 625 000	39 0625 0000	632456	15·8113883	6·2996052	4000000
251	6 30 01	15 813 251	39 6912 6001	631194	15·8429795	6·3079935	3984064
252	6 35 04	16 003 008	40 3275 8016	629941	15·8745079	6·3163596	3968254
253	6 40 09	16 194 277	40 9715 2081	628695	15·9059737	6·3247035	3952569
254	6 45 16	16 387 064	41 6231 4256	627456	15·9373775	6·3330255	3937008
255	6 50 25	16 581 375	42 2825 0625	626224	15·9687194	6·3413257	3921569
256	6 55 36	16 777 216	42 9496 7296	625000	16·0000000	6·3496042	3906250
257	6 60 49	16 974 593	43 6247 0401	623783	16·0312195	6·3578612	3891051
258	6 65 64	17 173 512	44 3076 6096	622573	16·0623784	6·3660968	3875969
259	6 70 81	17 373 979	44 9986 0561	621370	16·0934769	6·3743111	3861004
260	6 76 00	17 576 000	45 6976 0000	620174	16·1245155	6·3825043	3846154
261	6 81 21	17 779 581	46 4047 0641	618984	16·1554944	6·3906765	3831418
262	6 86 44	17 984 728	47 1199 8736	617802	16·1864141	6·3988279	3816794
263	6 91 69	18 191 447	47 8435 0561	616626	16·2172747	6·4069586	3802281
264	6 96 96	18 399 744	48 5753 2416	615457	16·2480768	6·4150687	3787879
265	7 02 25	18 609 625	49 3155 0625	614295	16·2788206	6·4231583	3773585
266	7 07 56	18 821 096	50 0641 1536	613139	16·3095064	6·4312276	3759398
267	7 12 89	19 034 163	50 8212 1521	611990	16·3401346	6·4392767	3745318
268	7 18 24	19 248 832	51 5868 6976	610847	16·3707055	6·4473057	3731343
269	7 23 61	19 465 109	52 3611 4321	609711	16·4012195	6·4553148	3717472
270	7 29 00	19 683 000	53 1441 0000	608581	16·4316767	6·4633041	3703704
271	7 34 41	19 902 511	53 9358 0481	607457	16·4620776	6·4712736	3690037
272	7 39 84	20 123 648	54 7363 2256	606339	16·4924225	6·4792236	3676471
273	7 45 29	20 346 417	55 5457 1841	605228	16·5227116	6·4871541	3663004
274	7 50 76	20 570 824	56 3640 5776	604122	16·5529454	6·4950653	3649635
275	7 56 25	20 796 875	57 1914 0625	603023	16·5831240	6·5029572	3636364
276	7 61 76	21 024 576	58 0278 2976	601929	16·6132477	6·5108301	3623188
277	7 67 29	21 253 933	58 8733 9441	600842	16·6433170	6·5186839	3610108
278	7 72 84	21 484 952	59 7281 6656	599760	16·6733320	6·5265189	3597122
279	7 78 41	21 717 639	60 5922 1281	598684	16·7032931	6·5343351	3584229
280	7 84 00	21 952 000	61 4656 0000	597614	16·7332005	6·5421326	3571429
281	7 89 61	22 188 041	62 3483 9521	596550	16·7630546	6·5499116	3558719
282	7 95 24	22 425 768	63 2406 6576	595491	16·7928556	6·5576722	3546099
283	8 00 89	22 665 187	64 1424 7921	594438	16·8226038	6·5654144	3533569
284	8 06 56	22 906 304	65 0539 0336	593391	16·8522995	6·5731385	3521127
285	8 12 25	23 149 125	65 9750 0625	592349	16·8819430	6·5808444	3508772
286	8 17 96	23 393 656	66 9058 5616	591312	16·9115345	6·5885323	3496503
287	8 23 69	23 639 903	67 8465 2161	590281	16·9410743	6·5962023	3484321
288	8 29 44	23 887 872	68 7970 7136	589256	16·9705627	6·6038545	3472222
289	8 35 21	24 137 569	69 7575 7441	588235	17·0000000	6·6114890	3460208
290	8 41 00	24 389 000	70 7281 0000	587220	17·0293864	6·6191059	3448276
291	8 46 81	24 642 .71	71 7087 1761	586210	17·0587221	6·6267054	3436426
292	8 52 64	24 897 088	72 6994 9696	585206	17·0880075	6·6342874	3424658
293	8 58 49	25 153 757	73 7005 0801	584206	17·1172428	6·6418522	3412969
294	8 64 36	25 412 184	74 7118 2096	583212	17·1464282	6·6493998	3401361
295	8 70 25	25 672 375	75 7335 0625	582223	17·1755640	6·6569302	3389831
296	8 76 16	25 934 336	76 7656 3456	581238	17·2046505	6·6644437	3378378
297	8 82 09	26 198 073	77 8082 7681	580259	17·2336879	6·6719403	3367003
298	8 88 04	26 463 592	78 8615 0416	579284	17·2626765	6·6794200	3355705
299	8 94 01	26 730 899	79 9253 8801	578315	17·2916165	6·6868831	3344482
300	9 00 00	27 000 000	81 0000 0000	577350	17·3205081	6·6943295	3333333

No. n	Square n^2	Cube n^3	Fourth power n^4	$\frac{1}{\sqrt{n}}$	Square root $\sqrt{n}$	Cube root $\sqrt[3]{n}$	Reciprocal $\frac{1}{n}$
				0·0			0·00
300	9 00 00	27 000 000	81 0000 0000	577350	17·3205081	6·6943295	3333333
301	9 06 01	27 270 901	82 0854 1201	576390	17·3493516	6·7017594	3322259
302	9 12 04	27 543 608	83 1816 9616	575435	17·3781472	6·7091729	3311258
303	9 18 09	27 818 127	84 2889 2481	574485	17·4068952	6·7165700	3300330
304	9 24 16	28 094 464	85 4071 7056	573539	17·4355958	6·7239508	3289474
305	9 30 25	28 372 625	86 5365 0625	572598	17·4642492	6·7313155	3278689
306	9 36 36	28 652 616	87 6770 0496	571662	17·4928557	6·7386641	3267974
307	9 42 49	28 934 443	88 8287 4001	570730	17·5214155	6·7459967	3257329
308	9 48 64	29 218 112	89 9917 8496	569803	17·5499288	6·7533134	3246753
309	9 54 81	29 503 629	91 1662 1361	568880	17·5783958	6·7606143	3236246
310	9 61 00	29 791 000	92 3521 0000	567962	17·6068169	6·7678995	3225806
311	9 67 21	30 080 231	93 5495 1841	567048	17·6351921	6·7751690	3215434
312	9 73 44	30 371 328	94 7585 4336	566139	17·6635217	6·7824229	3205128
313	9 79 69	30 664 297	95 9792 4961	565233	17·6918060	6·7896613	3194888
314	9 85 96	30 959 144	97 2117 1216	564333	17·7200451	6·7968844	3184713
315	9 92 25	31 255 875	98 4560 0625	563436	17·7482393	6·8040921	3174603
316	9 98 56	31 554 496	99 7122 0736	562544	17·7763888	6·8112846	3164557
317	10 04 89	31 855 013	100 9803 9121	561656	17·8044938	6·8184619	3154574
318	10 11 24	32 157 432	102 2606 3376	560772	17·8325545	6·8256242	3144654
319	10 17 61	32 461 759	103 5530 1121	559893	17·8605711	6·8327715	3134796
320	10 24 00	32 768 000	104 8576 0000	559017	17·8885438	6·8399038	3125000
321	10 30 41	33 076 161	106 1744 7681	558146	17·9164729	6·8470213	3115265
322	10 36 84	33 386 248	107 5037 1856	557278	17·9443584	6·8541240	3105590
323	10 43 29	33 698 267	108 8454 0241	556415	17·9722008	6·8612120	3095975
324	10 49 76	34 012 224	110 1996 0576	555556	18·0000000	6·8682855	3086420
325	10 56 25	34 328 125	111 5664 0625	554700	18·0277564	6·8753443	3076923
326	10 62 76	34 645 976	112 9458 8176	553849	18·0554701	6·8823888	3067485
327	10 69 29	34 965 783	114 3381 1041	553001	18·0831413	6·8894188	3058104
328	10 75 84	35 287 552	115 7431 7056	552158	18·1107703	6·8964345	3048780
329	10 82 41	35 611 289	117 1611 4081	551318	18·1383571	6·9034359	3039514
330	10 89 00	35 937 000	118 5921 0000	550482	18·1659021	6·9104232	3030303
331	10 95 61	36 264 691	120 0361 2721	549650	18·1934054	6·9173964	3021148
332	11 02 24	36 594 368	121 4933 0176	548821	18·2208672	6·9243556	3012048
333	11 08 89	36 926 037	122 9637 0321	547997	18·2482876	6·9313008	3003003
334	11 15 56	37 259 704	124 4474 1136	547176	18·2756669	6·9382321	2994012
335	11 22 25	37 595 375	125 9445 0625	546358	18·3030052	6·9451496	2985075
336	11 28 96	37 933 056	127 4550 6816	545545	18·3303028	6·9520533	2976190
337	11 35 69	38 272 753	128 9791 7761	544735	18·3575598	6·9589433	2967359
338	11 42 44	38 614 472	130 5169 1536	543928	18·3847763	6·9658198	2958580
339	11 49 21	38 958 219	132 0683 6241	543125	18·4119526	6·9726826	2949853
340	11 56 00	39 304 000	133 6336 0000	542326	18·4390889	6·9795320	2941176
341	11 62 81	39 651 821	135 2127 0961	541530	18·4661853	6·9863680	2932551
342	11 69 64	40 001 688	136 8057 7296	540738	18·4932420	6·9931907	2923977
343	11 76 49	40 353 607	138 4128 7201	539949	18·5202592	7·0000000	2915452
344	11 83 36	40 707 584	140 0340 8896	539164	18·5472370	7·0067961	2906977
345	11 90 25	41 063 625	141 6695 0625	538382	18·5741756	7·0135791	2898551
346	11 97 16	41 421 736	143 3192 0656	537603	18·6010752	7·0203490	2890173
347	12 04 09	41 781 923	144 9832 7281	536828	18·6279360	7·0271058	2881844
348	12 11 04	42 14[illegible] 192	146 6617 8816	536056	18·6547581	7·0338497	2873563
349	12 18 01	42 5[illegible] 549	148 3548 3601	535288	18·6815417	7·0405806	2865330
350	12 25 00	42 875 000	150 0625 0000	534522	18·7082869	7·0472987	2857143

No. n	Square n^2	Cube n^3	Fourth power n^4	$\frac{1}{\sqrt{n}}$	Square root $\sqrt{n}$	Cube root $\sqrt[3]{n}$	Reciprocal $\frac{1}{n}$
				0·0			0·00
350	12 25 00	42 875 000	150 0625 0000	534522	18·7082869	7·0472987	2857143
351	12 32 01	43 243 551	151 7848 6401	533761	18·7349940	7·0540041	2849003
352	12 39 04	43 614 208	153 5220 1216	533002	18·7616630	7·0606967	2840909
353	12 46 09	43 986 977	155 2740 2881	532246	18·7882942	7·0673766	2832861
354	12 53 16	44 361 864	157 0409 9856	531494	18·8148877	7·0740440	2824859
355	12 60 25	44 738 875	158 8230 0625	530745	18·8414437	7·0806988	2816901
356	12 67 36	45 118 016	160 6201 3696	529999	18·8679623	7·0873411	2808989
357	12 74 49	45 499 293	162 4324 7601	529256	18·8944436	7·0939709	2801120
358	12 81 64	45 882 712	164 2601 0896	528516	18·9208879	7·1005885	2793296
359	12 88 81	46 268 279	166 1031 2161	527780	18·9472953	7·1071937	2785515
360	12 96 00	46 656 000	167 9616 0000	527046	18·9736660	7·1137866	2777778
361	13 03 21	47 045 881	169 8356 3041	526316	19·0000000	7·1203674	2770083
362	13 10 44	47 437 928	171 7252 9936	525588	19·0262976	7·1269360	2762431
363	13 17 69	47 832 147	173 6306 9361	524864	19·0525589	7·1334925	2754821
364	13 24 96	48 228 544	175 5519 0016	524142	19·0787840	7·1400370	2747253
365	13 32 25	48 627 125	177 4890 0625	523424	19·1049732	7·1465695	2739726
366	13 39 56	49 027 896	179 4420 9936	522708	19·1311265	7·1530901	2732240
367	13 46 89	49 430 863	181 4112 6721	521996	19·1572441	7·1595988	2724796
368	13 54 24	49 836 032	183 3965 9776	521286	19·1833261	7·1660957	2717391
369	13 61 61	50 243 409	185 3981 7921	520579	19·2093727	7·1725809	2710027
370	13 69 00	50 653 000	187 4161 0000	519875	19·2353841	7·1790544	2702703
371	13 76 41	51 064 811	189 4504 4881	519174	19·2613603	7·1855162	2695418
372	13 83 84	51 478 848	191 5013 1456	518476	19·2873015	7·1919663	2688172
373	13 91 29	51 895 117	193 5687 8641	517780	19·3132079	7·1984050	2680965
374	13 98 76	52 313 624	195 6529 5376	517088	19·3390796	7·2048321	2673797
375	14 06 25	52 734 375	197 7539 0625	516398	19·3649167	7·2112479	2666667
376	14 13 76	53 157 376	199 8717 3376	515711	19·3907194	7·2176522	2659574
377	14 21 29	53 582 633	202 0065 2641	515026	19·4164878	7·2240451	2652520
378	14 28 84	54 010 152	204 1583 7456	514345	19·4422221	7·2304268	2645503
379	14 36 41	54 439 939	206 3273 6881	513665	19·4679223	7·2367972	2638522
380	14 44 00	54 872 000	208 5136 0000	512989	19·4935887	7·2431564	2631579
381	14 51 61	55 306 341	210 7171 5921	512316	19·5192213	7·2495045	2624672
382	14 59 24	55 742 968	212 9381 3776	511645	19·5448203	7·2558415	2617801
383	14 66 89	56 181 887	215 1766 2721	510976	19·5703858	7·2621674	2610966
384	14 74 56	56 623 104	217 4327 1936	510310	19·5959179	7·2684824	2604167
385	14 82 25	57 066 625	219 7065 0625	509647	19·6214169	7·2747863	2597403
386	14 89 96	57 512 456	221 9980 8016	508987	19·6468827	7·2810794	2590674
387	14 97 69	57 960 603	224 3075 3361	508329	19·6723156	7·2873616	2583979
388	15 05 44	58 411 072	226 6349 5936	507673	19·6977156	7·2936330	2577320
389	15 13 21	58 863 869	228 9804 5041	507020	19·7230829	7·2998937	2570694
390	15 21 00	59 319 000	231 3441 0000	506370	19·7484177	7·3061436	2564103
391	15 28 81	59 776 471	233 7260 0161	505722	19·7737199	7·3123828	2557545
392	15 36 64	60 236 288	236 1262 4896	505076	19·7989899	7·3186114	2551020
393	15 44 49	60 698 457	238 5449 3601	504433	19·8242276	7·3248294	2544529
394	15 52 36	61 162 984	240 9821 5696	503793	19·8494332	7·3310369	2538071
395	15 60 25	61 629 875	243 4380 0625	503155	19·8746069	7·3372339	2531646
396	15 68 16	62 099 136	245 9125 7856	502519	19·8997487	7·3434205	2525253
397	15 76 09	62 570 773	248 4059 6881	501886	19·9248588	7·3495966	2518892
398	15 84 04	63 044 792	250 9182 7216	501255	19·9499373	7·3557624	2512563
399	15 92 01	63 521 199	253 4495 8401	500626	19·9749844	7·3619178	2506266
400	16 00 00	64 000 000	256 0000 0000	500000	20·0000000	7·3680630	2500000

No. n	Square n^2	Cube n^3	Fourth power n^4	$\frac{1}{\sqrt{n}}$	Square root $\sqrt{n}$	Cube root $\sqrt[3]{n}$	Reciprocal $\frac{1}{n}$
				0·0			0·00
400	16 00 00	64 000 000	256 0000 0000	500000	20·0000000	7·3680630	2500000
401	16 08 01	64 481 201	258 5696 1601	499376	20·0249844	7·3741979	2493766
402	16 16 04	64 964 808	261 1585 2816	498755	20·0499377	7·3803227	2487562
403	16 24 09	65 450 827	263 7668 3281	498135	20·0748599	7·3864373	2481390
404	16 32 16	65 939 264	266 3946 2656	497519	20·0997512	7·3925418	2475248
405	16 40 25	66 430 125	269 0420 0625	496904	20·1246118	7·3986362	2469136
406	16 48 36	66 923 416	271 7090 6896	496292	20·1494417	7·4047206	2463054
407	16 56 49	67 419 143	274 3959 1201	495682	20·1742410	7·4107951	2457002
408	16 64 64	67 917 312	277 1026 3296	495074	20·1990099	7·4168595	2450980
409	16 72 81	68 417 929	279 8293 2961	494468	20·2237484	7·4229141	2444988
410	16 81 00	68 921 000	282 5761 0000	493865	20·2484567	7·4289588	2439024
411	16 89 21	69 426 531	285 3430 4241	493264	20·2731349	7·4349937	2433090
412	16 97 44	69 934 528	288 1302 5536	492665	20·2977831	7·4410189	2427184
413	17 05 69	70 444 997	290 9378 3761	492068	20·3224014	7·4470342	2421308
414	17 13 96	70 957 944	293 7658 8816	491473	20·3469899	7·4530399	2415459
415	17 22 25	71 473 375	296 6145 0625	490881	20·3715488	7·4590359	2409639
416	17 30 56	71 991 296	299 4837 9136	490290	20·3960781	7·4650223	2403846
417	17 38 89	72 511 713	302 3738 4321	489702	20·4205779	7·4709991	2398082
418	17 47 24	73 034 632	305 2847 6176	489116	20·4450483	7·4769664	2392344
419	17 55 61	73 560 059	308 2166 4721	488532	20·4694895	7·4829241	2386635
420	17 64 00	74 088 000	311 1696 0000	487950	20·4939015	7·4888724	2380952
421	17 72 41	74 618 461	314 1437 2081	487370	20·5182845	7·4948112	2375297
422	17 80 84	75 151 448	317 1391 1056	486792	20·5426386	7·5007407	2369668
423	17 89 29	75 686 967	320 1558 7041	486217	20·5669638	7·5066607	2364066
424	17 97 76	76 225 024	323 1941 0176	485643	20·5912603	7·5125715	2358491
425	18 06 25	76 765 625	326 2539 0625	485071	20·6155281	7·5184730	2352941
426	18 14 76	77 308 776	329 3353 8576	484502	20·6397674	7·5243652	2347418
427	18 23 29	77 854 483	332 4386 4241	483934	20·6639783	7·5302482	2341920
428	18 31 84	78 402 752	335 5637 7856	483368	20·6881609	7·5361220	2336449
429	18 40 41	78 953 589	338 7108 9681	482805	20·7123152	7·5419867	2331002
430	18 49 00	79 507 000	341 8801 0000	482243	20·7364414	7·5478423	2325581
431	18 57 61	80 062 991	345 0714 9121	481683	20·7605395	7·5536888	2320186
432	18 66 24	80 621 568	348 2851 7376	481125	20·7846097	7·5595263	2314815
433	18 74 89	81 182 737	351 5212 5121	480569	20·8086520	7·5653548	2309469
434	18 83 56	81 746 504	354 7798 2736	480015	20·8326667	7·5711743	2304147
435	18 92 25	82 312 875	358 0610 0625	479463	20·8566536	7·5769849	2298851
436	19 00 96	82 881 856	361 3648 9216	478913	20·8806130	7·5827865	2293578
437	19 09 69	83 453 453	364 6915 8961	478365	20·9045450	7·5885793	2288330
438	19 18 44	84 027 672	368 0412 0336	477818	20·9284495	7·5943633	2283105
439	19 27 21	84 604 519	371 4138 3841	477274	20·9523268	7·6001385	2277904
440	19 36 00	85 184 000	374 8096 0000	476731	20·9761770	7·6059049	2272727
441	19 44 81	85 766 121	378 2285 9361	476190	21·0000000	7·6116626	2267574
442	19 53 64	86 350 888	381 6709 2496	475651	21·0237960	7·6174116	2262443
443	19 62 49	86 938 307	385 1367 0001	475114	21·0475652	7·6231519	2257336
444	19 71 36	87 528 384	388 6260 2496	474579	21·0713075	7·6288836	2252252
445	19 80 25	88 121 125	392 1390 0625	474045	21·0950231	7·6346067	2247191
446	19 89 16	88 716 536	395 6757 5056	473514	21·1187121	7·6403212	2242152
447	19 98 09	89 314 623	399 2363 6481	472984	21·1423745	7·6460272	2237136
448	20 07 04	89 915 392	402 8209 5616	472456	21·1660105	7·6517247	223 143
449	20 16 01	90 518 849	406 4296 3201	471929	21·1896201	7·6574137	2227171
450	20 25 00	91 125 000	410 0625 0000	471405	21·2132034	7·6630943	2222222

No. n	Square n^2	Cube n^3	Fourth power n^4	$\frac{1}{\sqrt{n}}$	Square root $\sqrt{n}$	Cube root $\sqrt[3]{n}$	Reciprocal $\frac{1}{n}$
				0·0			0·00
450	20 25 00	91 125 000	410 0625 0000	471405	21·2132034	7·6630943	2222222
451	20 34 01	91 733 851	413 7196 6801	470882	21·2367606	7·6687665	2217295
452	20 43 04	92 345 408	417 4012 4416	470360	21·2602916	7·6744303	2212389
453	20 52 09	92 959 677	421 1073 3681	469841	21·2837967	7·6800857	2207506
454	20 61 16	93 576 664	424 8380 5456	469323	21·3072758	7·6857328	2202643
455	20 70 25	94 196 375	428 5935 0625	468807	21·3307290	7·6913717	2197802
456	20 79 36	94 818 816	432 3738 0096	468293	21·3541565	7·6970023	2192982
457	20 88 49	95 443 993	436 1790 4801	467780	21·3775583	7·7026246	2188184
458	20 97 64	96 071 912	440 0093 5696	467269	21·4009346	7·7082388	2183406
459	21 06 81	96 702 579	443 8648 3761	466760	21·4242853	7·7138448	2178649
460	21 16 00	97 336 000	447 7456 0000	466252	21·4476106	7·7194426	2173913
461	21 25 21	97 972 181	451 6517 5441	465746	21·4709106	7·7250324	2169197
462	21 34 44	98 611 128	455 5834 1136	465242	21·4941853	7·7306141	2164502
463	21 43 69	99 252 847	459 5406 8161	464739	21·5174348	7·7361877	2159827
464	21 52 96	99 897 344	463 5236 7616	464238	21·5406592	7·7417533	2155172
465	21 62 25	100 544 625	467 5325 0625	463739	21·5638587	7·7473109	2150538
466	21 71 56	101 194 696	471 5672 8336	463241	21·5870331	7·7528605	2145923
467	21 80 89	101 847 563	475 6281 1921	462745	21·6101828	7·7584023	2141328
468	21 90 24	102 503 232	479 7151 2576	462250	21·6333077	7·7639361	2136752
469	21 99 61	103 161 709	483 8284 1521	461757	21·6564078	7·7694620	2132196
470	22 09 00	103 823 000	487 9681 0000	461266	21·6794834	7·7749801	2127660
471	22 18 41	104 487 111	492 1342 9281	460776	21·7025344	7·7804904	2123142
472	22 27 84	105 154 048	496 3271 0656	460287	21·7255610	7·7859928	2118644
473	22 37 29	105 823 817	500 5466 5441	459800	21·7485632	7·7914875	2114165
474	22 46 76	106 496 424	504 7930 4976	459315	21·7715411	7·7969745	2109705
475	22 56 25	107 171 875	509 0664 0625	458831	21·7944947	7·8024538	2105263
476	22 65 76	107 850 176	513 3668 3776	458349	21·8174242	7·8079253	2100840
477	22 75 29	108 531 333	517 6944 5841	457869	21·8403297	7·8133892	2096436
478	22 84 84	109 215 352	522 0493 8256	457389	21·8632111	7·8188455	2092050
479	22 94 41	109 902 239	526 4317 2481	456912	21·8860686	7·8242942	2087683
480	23 04 00	110 592 000	530 8416 0000	456435	21·9089023	7·8297353	2083333
481	23 13 61	111 284 641	535 2791 2321	455961	21·9317122	7·8351688	2079002
482	23 23 24	111 980 168	539 7444 0976	455488	21·9544984	7·8405948	2074689
483	23 32 89	112 678 587	544 2375 7521	455016	21·9772610	7·8460134	2070393
484	23 42 56	113 379 904	548 7587 3536	454545	22·0000000	7·8514244	2066116
485	23 52 25	114 084 125	553 3080 0625	454077	22·0227155	7·8568280	2061856
486	23 61 96	114 791 256	557 8855 0416	453609	22·0454077	7·8622242	2057613
487	23 71 69	115 501 303	562 4913 4561	453143	22·0680765	7·8676130	2053388
488	23 81 44	116 214 272	567 1256 4736	452679	22·0907220	7·8729944	2049180
489	23 91 21	116 930 169	571 7885 2641	452216	22·1133444	7·8783684	2044990
490	24 01 00	117 649 000	576 4801 0000	451754	22·1359436	7·8837352	2040816
491	24 10 81	118 370 771	581 2004 8561	451294	22·1585198	7·8890946	2036660
492	24 20 64	119 095 488	585 9498 0096	450835	22·1810730	7·8944468	2032520
493	24 30 49	119 823 157	590 7281 6401	450377	22·2036033	7·8997917	2028398
494	24 40 36	120 553 784	595 5356 9296	449921	22·2261108	7·9051294	2024291
495	24 50 25	121 287 375	600 3725 0625	449467	22·2485955	7·9104599	2020202
496	24 60 16	122 023 936	605 2387 2256	449013	22·2710575	7·9157832	2016129
497	24 70 09	122 763 473	610 1344 6081	448561	22·2934968	7·9210994	2012072
498	24 80 04	123 505 992	615 0598 4016	448111	22·3159136	7·9264084	2008032
499	24 90 01	124 251 499	620 0149 8001	447661	22·3383079	7·9317104	2004008
500	25 00 00	125 000 000	625 0000 0000	447214	22·3606798	7·9370053	2000000

No. n	Square n^2	Cube n^3	Fourth power n^4	$\frac{1}{\sqrt{n}}$	Square root $\sqrt{n}$	Cube root $\sqrt[3]{n}$	Reciprocal $\frac{1}{n}$
				0·0			0·00
500	25 00 00	125 000 000	625 0000 0000	447214	22·3606798	7·9370053	2000000
501	25 10 01	125 751 501	630 0150 2001	446767	22·3830293	7·9422931	1996008
502	25 20 04	126 506 008	635 0601 6016	446322	22·4053565	7·9475739	1992032
503	25 30 09	127 263 527	640 1355 4081	445878	22·4276615	7·9528476	1988072
504	25 40 16	128 024 064	645 2412 8256	445435	22·4499443	7·9581144	1984127
505	25 50 25	128 787 625	650 3775 0625	444994	22·4722051	7·9633742	1980198
506	25 60 36	129 554 216	655 5443 3296	444554	22·4944438	7·9686271	1976285
507	25 70 49	130 323 843	660 7418 8401	444116	22·5166605	7·9738731	1972387
508	25 80 64	131 096 512	665 9702 8096	443678	22·5388553	7·9791122	1968504
509	25 90 81	131 872 229	671 2296 4561	443242	22·5610283	7·9843444	1964637
510	26 01 00	132 651 000	676 5201 0000	442807	22·5831796	7·9895697	1960784
511	26 11 21	133 432 831	681 8417 6641	442374	22·6053091	7·9947883	1956947
512	26 21 44	134 217 728	687 1947 6736	441942	22·6274170	8·0000000	1953125
513	26 31 69	135 005 697	692 5792 2561	441511	22·6495033	8·0052049	1949318
514	26 41 96	135 796 744	697 9952 6416	441081	22·6715681	8·0104031	1945525
515	26 52 25	136 590 875	703 4430 0625	440653	22·6936114	8·0155946	1941748
516	26 62 56	137 388 096	708 9225 7536	440225	22·7156334	8·0207793	1937984
517	26 72 89	138 188 413	714 4340 9521	439799	22·7376340	8·0259574	1934236
518	26 83 24	138 991 832	719 9776 8976	439375	22·7596134	8·0311287	1930502
519	26 93 61	139 798 359	725 5534 8321	438951	22·7815715	8·0362934	1926782
520	27 04 00	140 608 000	731 1616 0000	438529	22·8035085	8·0414515	1923077
521	27 14 41	141 420 761	736 8021 6481	438108	22·8254244	8·0466030	1919386
522	27 24 84	142 236 648	742 4753 0256	437688	22·8473193	8·0517479	1915709
523	27 35 29	143 055 667	748 1811 3841	437269	22·8691933	8·0568862	1912046
524	27 45 76	143 877 824	753 9197 9776	436852	22·8910463	8·0620180	1908397
525	27 56 25	144 703 125	759 6914 0625	436436	22·9128785	8·0671432	1904762
526	27 66 76	145 531 576	765 4960 8976	436021	22·9346899	8·0722620	1901141
527	27 77 29	146 363 183	771 3339 7441	435607	22·9564806	8·0773742	1897533
528	27 87 84	147 197 952	777 2051 8656	435194	22·9782506	8·0824800	1893939
529	27 98 41	148 035 889	783 1098 5281	434783	23·0000000	8·0875794	1890359
530	28 09 00	148 877 000	789 0481 0000	434372	23·0217289	8·0926723	1886792
531	28 19 61	149 721 291	795 0200 5521	433963	23·0434372	8·0977589	1883239
532	28 30 24	150 568 768	801 0258 4576	433555	23·0651252	8·1028390	1879699
533	28 40 89	151 419 437	807 0655 9921	433148	23·0867928	8·1079128	1876173
534	28 51 56	152 273 304	813 1394 4336	432742	23·1084400	8·1129803	1872659
535	28 62 25	153 130 375	819 2475 0625	432338	23·1300670	8·1180414	1869159
536	28 72 96	153 990 656	825 3899 1616	431934	23·1516738	8·1230962	1865672
537	28 83 69	154 854 153	831 5668 0161	431532	23·1732605	8·1281447	1862197
538	28 94 44	155 720 872	837 7782 9136	431131	23·1948270	8·1331870	1858736
539	29 05 21	156 590 819	844 0245 1441	430730	23·2163735	8·1382230	1855288
540	29 16 00	157 464 000	850 3056 0000	430331	23·2379001	8·1432528	1851852
541	29 26 81	158 340 421	856 6216 7761	429934	23·2594067	8·1482764	1848429
542	29 37 64	159 220 088	862 9728 7696	429537	23·2808935	8·1532939	1845018
543	29 48 49	160 103 007	869 3593 2801	429141	23·3023604	8·1583051	1841621
544	29 59 36	160 989 184	875 7811 6096	428746	23·3238076	8·1633102	1838235
545	29 70 25	161 878 625	882 2385 0625	428353	23·3452351	8·1683092	1834862
546	29 81 16	162 771 336	888 7314 9456	427960	23·3666429	8·1733020	1831502
547	29 92 09	163 667 323	895 2602 5681	427569	23·3880311	8·1782888	1828154
548	30 03 04	164 566 592	901 8249 2416	427179	23·4093998	8·1832695	1824818
549	30 14 01	165 469 149	908 4256 2801	426790	23·4307490	8·1882441	1821494
550	30 25 00	166 375 000	915 0625 0000	426401	23·4520788	8·1932127	1818182

No. n	Square n^2	Cube n^3	Fourth power n^4	$\frac{1}{\sqrt{n}}$	Square root $\sqrt{n}$	Cube root $\sqrt[3]{n}$	Reciprocal $\frac{1}{n}$
				0·0			0·00
550	30 25 00	166 375 000	915 0625 0000	426401	23·4520788	8·1932127	1818182
551	30 36 01	167 284 151	921 7356 7201	426014	23·4733892	8·1981753	1814882
552	30 47 04	168 196 608	928 4452 7616	425628	23·4946802	8·2031319	1811594
553	30 58 09	169 112 377	935 1914 4481	425243	23·5159520	8·2080825	1808318
554	30 69 16	170 031 464	941 9743 1056	424859	23·5372046	8·2130271	1805054
555	30 80 25	170 953 875	948 7940 0625	424476	23·5584380	8·2179657	1801802
556	30 91 36	171 879 616	955 6506 6496	424094	23·5796522	8·2228985	1798561
557	31 02 49	172 808 693	962 5444 2001	423714	23·6008474	8·2278254	1795332
558	31 13 64	173 741 112	969 4754 0496	423334	23·6220236	8·2327463	1792115
559	31 24 81	174 676 879	976 4437 5361	422955	23·6431808	8·2376614	1788909
560	31 36 00	175 616 000	983 4496 0000	422577	23·6643191	8·2425706	1785714
561	31 47 21	176 558 481	990 4930 7841	422200	23·6854386	8·2474740	1782531
562	31 58 44	177 504 328	997 5743 2336	421825	23·7065392	8·2523715	1779359
563	31 69 69	178 453 547	1004 6934 6961	421450	23·7276210	8·2572633	1776199
564	31 80 96	179 406 144	1011 8506 5216	421076	23·7486842	8·2621492	1773050
565	31 92 25	180 362 125	1019 0460 0625	420703	23·7697286	8·2670294	1769912
566	32 03 56	181 321 496	1026 2796 6736	420331	23·7907545	8·2719038	1766784
567	32 14 89	182 284 263	1033 5517 7121	419961	23·8117618	8·2767725	1763668
568	32 26 24	183 250 432	1040 8624 5376	419591	23·8327506	8·2816355	1760563
569	32 37 61	184 220 009	1048 2118 5121	419222	23·8537209	8·2864928	1757469
570	32 49 00	185 193 000	1055 6001 0000	418854	23·8746728	8·2913443	1754386
571	32 60 41	186 169 411	1063 0273 3681	418487	23·8956063	8·2961902	1751313
572	32 71 84	187 149 248	1070 4936 9856	418121	23·9165215	8·3010305	1748252
573	32 83 29	188 132 517	1077 9993 2241	417756	23·9374184	8·3058651	1745201
574	32 94 76	189 119 224	1085 5443 4576	417392	23·9582971	8·3106941	1742160
575	33 06 25	190 109 375	1093 1289 0625	417029	23·9791576	8·3155175	1739130
576	33 17 76	191 102 976	1100 7531 4176	416667	24·0000000	8·3203353	1736111
577	33 29 29	192 100 033	1108 4171 9041	416305	24·0208243	8·3251475	1733102
578	33 40 84	193 100 552	1116 1211 9056	415945	24·0416306	8·3299542	1730104
579	33 52 41	194 104 539	1123 8652 8081	415586	24·0624188	8·3347553	1727116
580	33 64 00	195 112 000	1131 6496 0000	415227	24·0831892	8·3395509	1724138
581	33 75 61	196 122 941	1139 4742 8721	414870	24·1039416	8·3443410	1721170
582	33 87 24	197 137 368	1147 3394 8176	414513	24·1246762	8·3491256	1718213
583	33 98 89	198 155 287	1155 2453 2321	414158	24·1453929	8·3539047	1715266
584	34 10 56	199 176 704	1163 1919 5136	413803	24·1660919	8·3586784	1712329
585	34 22 25	200 201 625	1171 1795 0625	413449	24·1867732	8·3634466	1709402
586	34 33 96	201 230 056	1179 2081 2816	413096	24·2074369	8·3682094	1706485
587	34 45 69	202 262 003	1187 2779 5761	412744	24·2280829	8·3729668	1703578
588	34 57 44	203 297 472	1195 3891 3536	412393	24·2487113	8·3777187	1700680
589	34 69 21	204 336 469	1203 5418 0241	412043	24·2693222	8·3824653	1697793
590	34 81 00	205 379 000	1211 7361 0000	411693	24·2899156	8·3872065	1694915
591	34 92 81	206 425 071	1219 9721 6961	411345	24·3104916	8·3919424	1692047
592	35 04 64	207 474 688	1228 2501 5296	410997	24·3310501	8·3966729	1689189
593	35 16 49	208 527 857	1236 5701 9201	410651	24·3515913	8·4013981	1686341
594	35 28 36	209 584 584	1244 9324 2896	410305	24·3721152	8·4061180	1683502
595	35 40 25	210 644 875	1253 3370 0625	409960	24·3926218	8·4108326	1680672
596	35 52 16	211 708 736	1261 7840 6656	409616	24·4131112	8·4155419	1677852
597	35 64 09	212 776 173	1270 2737 5281	409273	24·4335834	8·4202459	1675042
598	35 76 04	213 847 192	1278 8062 0816	408930	24·4540385	8·4249447	1672241
599	35 88 01	214 921 799	1287 3815 7601	408589	24·4744765	8·4296383	1669449
600	36 00 00	216 000 000	1296 0000 0000	408248	24·4948974	8·4343267	1666667

No. n	Square n^2	Cube n^3	Fourth power n^4	$\frac{1}{\sqrt{n}}$	Square root $\sqrt{n}$	Cube root $\sqrt[3]{n}$	Reciprocal $\frac{1}{n}$
				0·0			0·00
600	36 00 00	216 000 000	1296 0000 0000	408248	24·4948974	8·4343267	1666667
601	36 12 01	217 081 801	1304 6616 2401	407909	24·5153013	8·4390098	1663894
602	36 24 04	218 167 208	1313 3665 9216	407570	24·5356883	8·4436877	1661130
603	36 36 09	219 256 227	1322 1150 4881	407231	24·5560583	8·4483605	1658375
604	36 48 16	220 348 864	1330 9071 3856	406894	24·5764115	8·4530281	1655629
605	36 60 25	221 445 125	1339 7430 0625	406558	24·5967478	8·4576906	1652893
606	36 72 36	222 545 016	1348 6227 9696	406222	24·6170673	8·4623479	1650165
607	36 84 49	223 648 543	1357 5466 5601	405887	24·6373700	8·4670001	1647446
608	36 96 64	224 755 712	1366 5147 2896	405554	24·6576560	8·4716472	1644737
609	37 08 81	225 866 529	1375 5271 6161	405220	24·6779254	8·4762892	1642036
610	37 21 00	226 981 000	1384 5841 0000	404888	24·6981781	8·4809261	1639344
611	37 33 21	228 099 131	1393 6856 9041	404557	24·7184142	8·4855579	1636661
612	37 45 44	229 220 928	1402 8320 7936	404226	24·7386338	8·4901847	1633987
613	37 57 69	230 346 397	1412 0234 1361	403896	24·7588368	8·4948065	1631321
614	37 69 96	231 475 544	1421 2598 4016	403567	24·7790234	8·4994233	1628664
615	37 82 25	232 608 375	1430 5415 0625	403239	24·7991935	8·5040350	1626016
616	37 94 56	233 744 896	1439 8685 5936	402911	24·8193473	8·5086417	1623377
617	38 06 89	234 885 113	1449 2411 4721	402585	24·8394847	8·5132435	1620746
618	38 19 24	236 029 032	1458 6594 1776	402259	24·8596058	8·5178403	1618123
619	38 31 61	237 176 659	1468 1235 1921	401934	24·8797106	8·5224321	1615509
620	38 44 00	238 328 000	1477 6336 0000	401610	24·8997992	8·5270190	1612903
621	38 56 41	239 483 061	1487 1898 0881	401286	24·9198716	8·5316009	1610306
622	38 68 84	240 641 848	1496 7922 9456	400963	24·9399278	8·5361780	1607717
623	38 81 29	241 804 367	1506 4412 0641	400642	24·9599679	8·5407501	1605136
624	38 93 76	242 970 624	1516 1366 9376	400320	24·9799920	8·5453174	1602564
625	39 06 25	244 140 625	1525 8789 0625	400000	25·0000000	8·5498797	1600000
626	39 18 76	245 314 376	1535 6679 9376	399680	25·0199920	8·5544372	1597444
627	39 31 29	246 491 883	1545 5041 0641	399362	25·0399681	8·5589899	1594896
628	39 43 84	247 673 152	1555 3873 9456	399043	25·0599282	8·5635377	1592357
629	39 56 41	248 858 189	1565 3180 0881	398726	25·0798724	8·5680807	1589825
630	39 69 00	250 047 000	1575 2961 0000	398410	25·0998008	8·5726189	1587302
631	39 81 61	251 239 591	1585 3218 1921	398094	25·1197134	8·5771523	1584786
632	39 94 24	252 435 968	1595 3953 1776	397779	25·1396102	8·5816809	1582278
633	40 06 89	253 636 137	1605 5167 4721	397464	25·1594913	8·5862047	1579779
634	40 19 56	254 840 104	1615 6862 5936	397151	25·1793566	8·5907237	1577287
635	40 32 25	256 047 875	1625 9040 0625	396838	25·1992063	8·5952380	1574803
636	40 44 96	257 259 456	1636 1701 4016	396526	25·2190404	8·5997476	1572327
637	40 57 69	258 474 853	1646 4848 1361	396214	25·2388589	8·6042524	1569859
638	40 70 44	259 694 072	1656 8481 7936	395904	25·2586619	8·6087526	1567398
639	40 83 21	260 917 119	1667 2603 9041	395594	25·2784493	8·6132480	1564945
640	40 96 00	262 144 000	1677 7216 0000	395285	25·2982213	8·6177388	1562500
641	41 08 81	263 374 721	1688 2319 6161	394976	25·3179778	8·6222248	1560062
642	41 21 64	264 609 288	1698 7916 2896	394669	25·3377189	8·6267062	1557632
643	41 34 49	265 847 707	1709 4007 5601	394362	25·3574447	8·6311830	1555210
644	41 47 36	267 089 984	1720 0594 9696	394055	25·3771551	8·6356551	1552795
645	41 60 25	268 336 125	1730 7680 0625	393750	25·3968502	8·6401226	1550388
646	41 73 16	269 586 136	1741 5264 3856	393445	25·4165301	8·6445855	1547988
647	41 86 09	270 840 023	1752 3349 4881	393141	25·4361947	8·6490437	1545595
648	41 99 04	272 097 792	1763 1936 9216	392837	25·4558441	8·6534974	1543210
649	42 12 01	273 359 449	1774 1028 2401	392534	25·4754784	8·6579465	1540832
650	42 25 00	274 625 000	1785 0625 0000	392232	25·4950976	8·6623911	1538462

No. n	Square n^2	Cube n^3	Fourth power n^4	$\frac{1}{\sqrt{n}}$	Square root $\sqrt{n}$	Cube root $\sqrt[3]{n}$	Reciprocal $\frac{1}{n}$
				0·0			0·00
650	42 25 00	274 625 000	1785 0625 0000	392232	25·4950976	8·6623911	1538462
651	42 38 01	275 894 451	1796 0728 7601	391931	25·5147016	8·6668310	1536098
652	42 51 04	277 167 808	1807 1341 0816	391630	25·5342907	8·6712665	1533742
653	42 64 09	278 445 077	1818 2463 5281	391330	25·5538647	8·6756974	1531394
654	42 77 16	279 726 264	1829 4097 6656	391031	25·5734237	8·6801237	1529052
655	42 90 25	281 011 375	1840 6245 0625	390732	25·5929678	8·6845456	1526718
656	43 03 36	282 300 416	1851 8907 2896	390434	25·6124969	8·6889630	1524390
657	43 16 49	283 593 393	1863 2085 9201	390137	25·6320112	8·6933759	1522070
658	43 29 64	284 890 312	1874 5782 5296	389841	25·6515107	8·6977843	1519757
659	43 42 81	286 191 179	1885 9998 6961	389545	25·6709953	8·7021882	1517451
660	43 56 00	287 496 000	1897 4736 0000	389249	25·6904652	8·7065877	1515152
661	43 69 21	288 804 781	1908 9996 0241	388955	25·7099203	8·7109827	1512859
662	43 82 44	290 117 528	1920 5780 3536	388661	25·7293607	8·7153734	1510574
663	43 95 69	291 434 247	1932 2090 5761	388368	25·7487864	8·7197596	1508296
664	44 08 96	292 754 944	1943 8928 2816	388075	25·7681975	8·7241413	1506024
665	44 22 25	294 079 625	1955 6295 0625	387783	25·7875939	8·7285187	1503759
666	44 35 56	295 408 296	1967 4192 5136	387492	25·8069758	8·7328917	1501502
667	44 48 89	296 740 963	1979 2622 2321	387202	25·8263431	8·7372604	1499250
668	44 62 24	298 077 632	1991 1585 8176	386912	25·8456960	8·7416246	1497006
669	44 75 61	299 418 309	2003 1084 8721	386622	25·8650343	8·7459846	1494768
670	44 89 00	300 763 000	2015 1121 0000	386334	25·8843582	8·7503401	1492537
671	45 02 41	302 111 711	2027 1695 8081	386046	25·9036677	8·7546914	1490313
672	45 15 84	303 464 448	2039 2810 9056	385758	25·9229628	8·7590383	1488095
673	45 29 29	304 821 217	2051 4467 9041	385472	25·9422435	8·7633809	1485884
674	45 42 76	306 182 024	2063 6668 4176	385186	25 9615100	8·7677192	1483680
675	45 56 25	307 546 875	2075 9414 0625	384900	25·9807621	8·7720532	1481481
676	45 69 76	308 915 776	2088 2706 4576	304015	26·0000000	8·7763830	1479290
677	45 83 29	310 288 733	2100 6547 2241	384331	26·0192237	8·7807084	1477105
678	45 96 84	311 665 752	2113 0937 9856	384048	26·0384331	8·7850296	1474926
679	46 10 41	313 046 839	2125 5880 3681	383765	26·0576284	8·7893466	1472754
680	46 24 00	314 432 900	2138 1376 0000	383482	26·0768096	8·7936593	1470588
681	46 37 61	315 821 241	2150 7426 5121	383201	26·0959767	8·7979679	1468429
682	46 51 24	317 214 568	2163 4033 5376	382920	26·1151297	8·8022721	1466276
683	46 64 89	318 611 987	2176 1198 7121	382639	26·1342687	8·8065722	1464129
684	46 78 56	320 013 504	2188 8923 6736	382360	26·1533937	8·8108681	1461988
685	46 92 25	321 419 125	2201 7210 0625	382080	26·1725047	8·8151598	1459854
686	47 05 96	322 828 856	2214 6059 5216	381802	26·1916017	8·8194473	1457726
687	47 19 69	324 242 703	2227 5473 6961	381524	26·2106848	8·8237307	1455604
688	47 33 44	325 660 672	2240 5454 2336	381246	26·2297541	8·8280099	1453488
689	47 47 21	327 082 769	2253 6002 7841	380970	26·2488095	8·8322850	1451379
690	47 61 00	328 509 000	2266 7121 0000	380693	26·2678511	8·8365559	1449275
691	47 74 81	329 939 371	2279 8810 5361	380418	26·2868789	8·8408227	1447178
692	47 88 64	331 373 888	2293 1073 0496	380143	26·3058929	8·8450854	1445087
693	48 02 49	332 812 557	2306 3910 2001	379869	26·3248932	8·8493440	1443001
694	48 16 36	334 255 384	2319 7323 6496	379595	26·3438797	8·8535985	1440922
695	48 30 25	335 702 375	2333 1315 0625	379322	26·3628527	8·8578489	1438849
696	48 44 16	337 153 536	2346 5886 1056	379049	26·3818119	8·8620952	1436782
697	48 58 09	338 608 873	2360 1038 4481	378777	26·4007576	8·8663375	1434720
698	48 72 04	340 068 392	2373 6773 7616	378506	26·4196896	8·8705757	1432665
699	48 86 01	341 532 099	2387 3093 7201	378235	26·4386081	8·8748099	1430615
700	49 00 00	343 000 000	2401 0000 0000	377964	26·4575131	8·8790400	1428571

No. n	Square n^2	Cube n^3	Fourth power n^4	$\frac{1}{\sqrt{n}}$	Square root $\sqrt{n}$	Cube root $\sqrt[3]{n}$	Reciprocal $\frac{1}{n}$
				0·0			0·00
700	49 00 00	343 000 000	2401 0000 0000	377964	26·4575131	8·8790400	1428571
701	49 14 01	344 472 101	2414 7494 2801	377695	26·4764046	8·8832661	1426534
702	49 28 04	345 948 408	2428 5578 2416	377426	26·4952826	8·8874882	1424501
703	49 42 09	347 428 927	2442 4253 5681	377157	26·5141472	8·8917063	1422475
704	49 56 16	348 913 664	2456 3521 9456	376889	26·5329983	8·8959204	1420455
705	49 70 25	350 402 625	2470 3385 0625	376622	26·5518361	8·9001305	1418440
706	49 84 36	351 895 816	2484 3844 6096	376355	26·5706605	8·9043366	1416431
707	49 98 49	353 393 243	2498 4902 2801	376089	26·5894716	8·9085387	1414427
708	50 12 64	354 894 912	2512 6559 7696	375823	26·6082694	8·9127369	1412429
709	50 26 81	356 400 829	2526 8818 7761	375558	26·6270539	8·9169311	1410437
710	50 41 00	357 911 000	2541 1681 0000	375293	26·6458252	8·9211214	1408451
711	50 55 21	359 425 431	2555 5148 1441	375029	26·6645833	8·9253078	1406470
712	50 69 44	360 944 128	2569 9221 9136	374766	26·6833281	8·9294902	1404494
713	50 83 69	362 467 097	2584 3904 0161	374503	26·7020598	8·9336687	1402525
714	50 97 96	363 994 344	2598 9196 1616	374241	26·7207784	8·9378433	1400560
715	51 12 25	365 525 875	2613 5100 0625	373979	26·7394839	8·9420140	1398601
716	51 26 56	367 061 696	2628 1617 4336	373718	26·7581763	8·9461809	1396648
717	51 40 89	368 601 813	2642 8749 9921	373457	26·7768557	8·9503438	1394700
718	51 55 24	370 146 232	2657 6499 4576	373197	26·7955220	8 9545029	1392758
719	51 69 61	371 694 959	2672 4867 5521	372937	26·8141754	8·9586581	1390821
720	51 84 00	373 248 000	2687 3856 0000	372678	26·8328157	8·9628095	1388889
721	51 98 41	374 805 361	2702 3466 5281	372419	26·8514432	8·9669570	1386963
722	52 12 84	376 367 048	2717 3700 8656	372161	26·8700577	8·9711007	1385042
723	52 27 29	377 933 067	2732 4560 7441	371904	26·8886593	8·9752406	1383126
724	52 41 76	379 503 424	2747 6047 8976	371647	26·9072481	8·9793766	1381215
725	52 56 25	381 078 125	2762 8164 0625	371391	26·9258240	8·9835089	1379310
726	52 70 76	382 657 176	2778 0910 9776	371135	26·9443872	8·9876373	1377410
727	52 85 29	384 240 583	2793 4290 3841	370879	26·9629375	8·9917620	1375516
728	52 99 84	385 828 352	2808 8304 0256	370625	26·9814751	8·9958829	1373626
729	53 14 41	387 420 489	2824 2953 6481	370370	27·0000000	9·0000000	1371742
730	53 29 00	389 017 000	2839 8241 0000	370117	27·0185122	9·0041133	1369863
731	53 43 61	390 617 891	2855 4167 8321	369863	27·0370117	9·0082229	1367989
732	53 58 24	392 223 168	2871 0735 8976	369611	27·0554985	9·0123288	1366120
733	53 72 89	393 832 837	2886 7946 9521	369358	27·0739727	9·0164309	1364256
734	53 87 56	395 446 904	2902 5802 7536	369107	27·0924344	9·0205293	1362398
735	54 02 25	397 065 375	2918 4305 0625	368856	27·1108834	9·0246239	1360544
736	54 16 96	398 688 256	2934 3455 6416	368605	27·1293199	9·0287149	1358696
737	54 31 69	400 315 553	2950 3256 2561	368355	27·1477439	9·0328021	1356852
738	54 46 44	401 947 272	2966 3708 6736	368105	27·1661554	9·0368857	1355014
739	54 61 21	403 583 419	2982 4814 6641	367856	27·1845544	9·0409655	1353180
740	54 76 00	405 224 000	2998 6576 0000	367607	27·2029410	9·0450417	1351351
741	54 90 81	406 869 021	3014 8994 4561	367359	27·2213152	9·0491142	1349528
742	55 05 64	408 518 488	3031 2071 8096	367112	27·2396769	9·0531831	1347709
743	55 20 49	410 172 407	3047 5809 8401	366864	27·2580263	9·0572482	1345895
744	55 35 36	411 830 784	3064 0210 3296	366618	27·2763634	9·0613098	1344086
745	55 50 25	413 493 625	3080 5275 0625	366372	27·2946881	9·0653677	1342282
746	55 65 16	415 160 936	3097 1005 8256	366126	27·3130006	9·0694220	1340483
747	55 80 09	416 832 723	3113 7404 4081	365881	27·3313007	9·0734726	1338688
748	55 95 04	418 508 992	3130 4472 6016	365636	27·3495887	9·0775197	1336898
749	56 10 01	420 189 749	3147 2212 2001	365392	27·3678644	9·0815631	1335113
750	56 25 00	421 875 000	3164 0625 0000	365148	27·3861279	9·0856030	1333333

No. n	Square n^2	Cube n^3	Fourth power n^4	$\frac{1}{\sqrt{n}}$	Square root $\sqrt{n}$	Cube root $\sqrt[3]{n}$	Reciprocal $\frac{1}{n}$
				0·0			0·00
750	56 25 00	421 875 000	3164 0625 0000	365148	27·3861279	9·0856030	1333333
751	56 40 01	423 564 751	3180 9712 8001	364905	27·4043792	9·0896392	1331558
752	56 55 04	425 259 008	3197 9477 4016	364662	27·4226184	9·0936719	1329787
753	56 70 09	426 957 777	3214 9920 6081	364420	27·4408455	9·0977010	1328021
754	56 85 16	428 661 064	3232 1044 2256	364179	27·4590604	9·1017265	1326260
755	57 00 25	430 368 875	3249 2850 0625	363937	27·4772633	9·1057485	1324503
756	57 15 36	432 081 216	3266 5339 9296	363696	27·4954542	9·1097669	1322751
757	57 30 49	433 798 093	3283 8515 6401	363456	27·5136330	9·1137818	1321004
758	57 45 64	435 519 512	3301 2379 0096	363216	27·5317998	9·1177931	1319261
759	57 60 81	437 245 479	3318 6931 8561	362977	27·5499546	9·1218010	1317523
760	57 76 00	438 976 000	3336 2176 0000	362738	27·5680975	9·1258053	1315789
761	57 91 21	440 711 081	3353 8113 2641	362500	27·5862284	9·1298061	1314060
762	58 06 44	442 450 728	3371 4745 4736	362262	27·6043475	9·1338034	1312336
763	58 21 69	444 194 947	3389 2074 4561	362024	27·6224546	9·1377971	1310616
764	58 36 96	445 943 744	3407 0102 0416	361787	27·6405499	9·1417874	1308901
765	58 52 25	447 697 125	3424 8830 0625	361551	27·6586334	9·1457743	1307190
766	58 67 56	449 455 096	3442 8260 3536	361315	27·6767050	9·1497576	1305483
767	58 82 89	451 217 663	3460 8394 7521	361079	27·6947648	9·1537375	1303781
768	58 98 24	452 984 832	3478 9235 0976	360844	27·7128129	9·1577139	1302083
769	59 13 61	454 756 609	3497 0783 2321	360609	27·7308492	9·1616869	1300390
770	59 29 00	456 533 000	3515 3041 0000	360375	27·7488739	9·1656565	1298701
771	59 44 41	458 314 011	3533 6010 2481	360141	27·7668868	9·1696226	1297017
772	59 59 84	460 099 648	3551 9692 8256	359908	27·7848880	9·1735852	1295337
773	59 75 29	461 889 917	3570 4090 5841	359675	27·8028775	9·1775445	1293661
774	59 90 76	463 684 824	3588 9205 3776	359443	27·8208555	9·1815003	1291990
775	60 06 25	465 484 375	3607 5039 0625	359211	27·8388218	9·1854528	1290323
776	60 21 76	467 288 576	3626 1593 4976	358979	27·8567766	9·1894018	1288660
777	60 37 29	469 097 433	3644 8870 5441	358748	27·8747197	9·1933474	1287001
778	60 52 84	470 910 952	3663 6872 0656	358517	27·8926514	9·1972897	1285347
779	60 68 41	472 729 139	3682 5599 9281	358287	27·9105715	9·2012286	1283697
780	60 84 00	474 552 000	3701 5056 0000	358057	27·9284801	9·2051641	1282051
781	60 99 61	476 379 541	3720 5242 1521	357828	27·9463772	9·2090962	1280410
782	61 15 24	478 211 768	3739 6160 2576	357599	27·9642629	9·2130250	1278772
783	61 30 89	480 048 687	3758 7812 1921	357371	27·9821372	9·2169505	1277139
784	61 46 56	481 890 304	3778 0199 8336	357143	28·0000000	9·2208726	1275510
785	61 62 25	483 736 625	3797 3325 0625	356915	28·0178515	9·2247914	1273885
786	61 77 96	485 587 656	3816 7189 7616	356688	28·0356915	9·2287068	1272265
787	61 93 69	487 443 403	3836 1795 8161	356462	28·0535203	9·2326189	1270648
788	62 09 44	489 303 872	3855 7145 1136	356235	28·0713377	9·2365277	1269036
789	62 25 21	491 169 069	3875 3239 5441	356009	28·0891438	9·2404333	1267427
790	62 41 00	493 039 000	3895 0081 0000	355784	28·1069386	9·2443355	1265823
791	62 56 81	494 913 671	3914 7671 3761	355559	28·1247222	9·2482344	1264223
792	62 72 64	496 793 088	3934 6012 5696	355335	28·1424946	9·2521300	1262626
793	62 88 49	498 677 257	3954 5106 4801	355110	28·1602557	9·2560224	1261034
794	63 04 36	500 566 184	3974 4955 0096	354887	28·1780056	9·2599115	1259446
795	63 20 25	502 459 875	3994 5560 0625	354663	28·1957444	9·2637973	1257862
796	63 36 16	504 358 336	4014 6923 5456	354441	28·2134720	9·2676798	1256281
797	63 52 09	506 261 573	4034 9047 3681	354218	28·2311884	9·2715592	1254705
798	63 68 04	508 169 592	4055 1933 4416	353996	28·2488938	9·2754352	1253133
799	63 84 01	510 082 399	4075 5583 6801	353775	28·2665881	9·2793081	1251564
800	64 00 00	512 000 000	4096 0000 0000	353553	28·2842712	9·2831777	1250000

No. n	Square n^2	Cube n^3	Fourth power n^4	$\frac{1}{\sqrt{n}}$	Square root $\sqrt{n}$	Cube root $\sqrt[3]{n}$	Reciprocal $\frac{1}{n}$
				0·0			0·00
800	64 00 00	512 000 000	4096 0000 0000	353553	28·2842712	9·2831777	1250000
801	64 16 01	513 922 401	4116 5184 3201	353333	28·3019434	9·2870440	1248439
802	64 32 04	515 849 608	4137 1138 5616	353112	28·3196045	9·2909072	1246883
803	64 48 09	517 781 627	4157 7864 6481	352892	28·3372546	9·2947672	1245330
804	64 64 16	519 718 464	4178 5364 5056	352673	28·3548938	9·2986239	1243781
805	64 80 25	521 660 125	4199 3640 0625	352454	28·3725219	9·3024775	1242236
806	64 96 36	523 606 616	4220 2693 2496	352235	28·3901391	9·3063278	1240695
807	65 12 49	525 557 943	4241 2526 0001	352017	28·4077454	9·3101750	1239157
808	65 28 64	527 514 112	4262 3140 2496	351799	28·4253408	9·3140190	1237624
809	65 44 81	529 475 129	4283 4537 9361	351581	28·4429253	9·3178598	1236094
810	65 61 00	531 441 000	4304 6721 0000	351364	28·4604989	9·3216975	1234568
811	65 77 21	533 411 731	4325 9691 3841	351147	28·4780617	9·3255320	1233046
812	65 93 44	535 387 328	4347 3451 0336	350931	28·4956137	9·3293634	1231527
813	66 09 69	537 367 797	4368 8001 8961	350715	28·5131549	9·3331916	1230012
814	66 25 96	539 353 144	4390 3345 9216	350500	28·5306852	9·3370167	1228501
815	66 42 25	541 343 375	4411 9485 0625	350285	28·5482048	9·3408386	1226994
816	66 58 56	543 338 496	4433 6421 2736	350070	28·5657137	9·3446575	1225490
817	66 74 89	545 338 513	4455 4156 5121	349856	28·5832119	9·3484732	1223990
818	66 91 24	547 343 432	4477 2692 7376	349642	28·6006993	9·3522858	1222494
819	67 07 61	549 353 259	4499 2031 9121	349428	28·6181760	9·3560952	1221001
820	67 24 00	551 368 000	4521 2176 0000	349215	28·6356421	9·3599016	1219512
821	67 40 41	553 387 661	4543 3126 9681	349002	28·6530976	9·3637049	1218027
822	67 56 84	555 412 248	4565 4886 7856	348790	28·6705424	9·3675051	1216545
823	67 73 29	557 441 767	4587 7457 4241	348578	28·6879766	9·3713022	1215067
824	67 89 76	559 476 224	4610 0840 8576	348367	28·7054002	9·3750963	1213592
825	68 06 25	561 515 625	4632 5039 0625	348155	28·7228132	9·3788873	1212121
826	68 22 76	563 559 976	4655 0054 0176	347945	28·7402157	9·3826752	1210654
827	68 39 29	565 609 283	4677 5887 7041	347734	28·7576077	9·3864601	1209190
828	68 55 84	567 663 552	4700 2542 1056	347524	28·7749891	9·3902419	1207729
829	68 72 41	569 722 789	4723 0019 2081	347314	28·7923601	9·3940206	1206273
830	68 89 00	571 787 000	4745 8321 0000	347105	28·8097206	9·3977964	1204819
831	69 05 61	573 856 191	4768 7449 4721	346896	28·8270706	9·4015691	1203369
832	69 22 24	575 930 368	4791 7406 6176	346688	28·8444102	9·4053388	1201923
833	69 38 89	578 009 537	4814 8194 4321	346479	28·8617394	9·4091054	1200480
834	69 55 56	580 093 704	4837 9814 9136	346272	28·8790582	9·4128690	1199041
835	69 72 25	582 182 875	4861 2270 0625	346064	28·8963666	9·4166297	1197605
836	69 88 96	584 277 056	4884 5561 8816	345857	28·9136646	9·4203873	1196172
837	70 05 69	586 376 253	4907 9692 3761	345651	28·9309523	9·4241420	1194743
838	70 22 44	588 480 472	4931 4663 5536	345444	28·9482297	9·4278936	1193317
839	70 39 21	590 589 719	4955 0477 4241	345238	28·9654967	9·4316423	1191895
840	70 56 00	592 704 000	4978 7136 0000	345033	28·9827535	9·4353880	1190476
841	70 72 81	594 823 321	5002 4641 2961	344828	29·0000000	9·4391307	1189061
842	70 89 64	596 947 688	5026 2995 3296	344623	29·0172363	9·4428704	1187648
843	71 06 49	599 077 107	5050 2200 1201	344418	29·0344623	9·4466072	1186240
844	71 23 36	601 211 584	5074 2257 6896	344214	29·0516781	9·4503411	1184834
845	71 40 25	603 351 125	5098 3170 0625	344010	29·0688837	9·4540719	1183432
846	71 57 16	605 495 736	5122 4939 2656	343807	29·0860791	9·4577999	1182033
847	71 74 09	607 645 423	5146 7567 3281	343604	29·1032644	9·4615249	1180638
848	71 91 04	609 800 192	5171 1056 2816	343401	29·1204396	9·4652470	1179245
849	72 08 01	611 960 049	5195 5408 1601	343199	29·1376046	9·4689661	1177856
850	72 25 00	614 125 000	5220 0625 0000	342997	29·1547595	9·4726824	1176471

No. n	Square n^2	Cube n^3	Fourth power n^4	$\frac{1}{\sqrt{n}}$	Square root $\sqrt{n}$	Cube root $\sqrt[3]{n}$	Reciprocal $\frac{1}{n}$
				0·0			0·00
850	72 25 00	614 125 000	5220 0625 0000	342997	29·1547595	9·4726824	1176471
851	72 42 01	616 295 051	5244 6708 8401	342796	29·1719043	9·4763957	1175088
852	72 59 04	618 470 208	5269 3661 7216	342594	29·1890390	9·4801061	1173709
853	72 76 09	620 650 477	5294 1485 6881	342393	29·2061637	9·4838136	1172333
854	72 93 16	622 835 864	5319 0182 7856	342193	29·2232784	9·4875182	1170960
855	73 10 25	625 026 375	5343 9755 0625	341993	29·2403830	9·4912200	1169591
856	73 27 36	627 222 016	5369 0204 5696	341793	29·2574777	9·4949188	1168224
857	73 44 49	629 422 793	5394 1533 3601	341593	29·2745623	9·4986148	1166861
858	73 61 64	631 628 712	5419 3743 4896	341394	29·2916370	9·5023078	1165501
859	73 78 81	633 839 779	5444 6837 0161	341196	29·3087018	9·5059981	1164144
860	73 96 00	636 056 000	5470 0816 0000	340997	29·3257566	9·5096854	1162791
861	74 13 21	638 277 381	5495 5682 5041	340799	29·3428015	9·5133699	1161440
862	74 30 44	640 503 928	5521 1438 5936	340601	29·3598365	9·5170516	1160093
863	74 47 69	642 735 647	5546 8086 3361	340404	29·3768616	9·5207304	1158749
864	74 64 96	644 972 544	5572 5627 8016	340207	29·3938769	9·5244063	1157407
865	74 82 25	647 214 625	5598 4065 0625	340010	29·4108823	9·5280794	1156069
866	74 99 56	649 461 896	5624 3400 1936	339814	29·4278779	9·5317497	1154734
867	75 16 89	651 714 363	5650 3635 2721	339618	29·4448637	9·5354172	1153403
868	75 34 24	653 972 032	5676 4772 3776	339422	29·4618397	9·5390818	1152074
869	75 51 61	656 234 909	5702 6813 5921	339227	29·4788059	9·5427437	1150748
870	75 69 00	658 503 000	5728 9761 0000	339032	29·4957624	9·5464027	1149425
871	75 86 41	660 776 311	5755 3616 6881	338837	29·5127091	9·5500589	1148106
872	76 03 84	663 054 848	5781 8382 7456	338643	29·5296461	9·5537124	1146789
873	76 21 29	665 338 617	5808 4061 2641	338449	29·5465734	9·5573630	1145475
874	76 38 76	667 627 624	5835 0654 3376	338255	29·5634910	9·5610108	1144165
875	76 56 25	669 921 875	5861 8164 0625	338062	29·5803989	9·5646559	1142857
876	76 73 76	672 221 376	5888 6592 5376	337869	29·5972972	9·5682982	1141553
877	76 91 29	674 526 133	5915 5941 8641	337676	29·6141858	9·5719377	1140251
878	77 08 84	676 836 152	5942 6214 1456	337484	29·6310648	9·5755745	1138952
879	77 26 41	679 151 439	5969 7411 4881	337292	29·6479342	9·5792085	1137656
880	77 44 00	681 472 000	5996 9536 0000	337100	29·6647939	9·5828397	1136364
881	77 61 61	683 797 841	6024 2589 7921	336909	29·6816442	9·5864682	1135074
882	77 79 24	686 128 968	6051 6574 9776	336718	29·6984848	9·5900939	1133787
883	77 96 89	688 465 387	6079 1493 6721	336527	29·7153159	9·5937170	1132503
884	78 14 56	690 807 104	6106 7347 9936	336336	29·7321375	9·5973372	1131222
885	78 32 25	693 154 125	6134 4140 0625	336146	29·7489496	9·6009548	1129944
886	78 49 96	695 506 456	6162 1872 0016	335957	29·7657521	9·6045696	1128668
887	78 67 69	697 864 103	6190 0545 9361	335767	29·7825452	9·6081817	1127396
888	78 85 44	700 227 072	6218 0163 9936	335578	29·7993289	9·6117911	1126126
889	79 03 21	702 595 369	6246 0728 3041	335389	29·8161030	9·6153977	1124859
890	79 21 00	704 969 000	6274 2241 0000	335201	29·8328678	9·6190017	1123596
891	79 38 81	707 347 971	6302 4704 2161	335013	29·8496231	9·6226030	1122334
892	79 56 64	709 732 288	6330 8120 0896	334825	29·8663690	9·6262016	1121076
893	79 74 49	712 121 957	6359 2490 7601	334637	29·8831056	9·6297975	1119821
894	79 92 36	714 516 984	6387 7818 3696	334450	29·8998328	9·6333907	1118568
895	80 10 25	716 917 375	6416 4105 0625	334263	29·9165506	9·6369812	1117318
896	80 28 16	719 323 136	6445 1352 9856	334077	29·9332591	9·6405691	1116071
897	80 46 09	721 734 273	6473 9564 2881	333890	29·9499583	9·6441542	1114827
898	80 64 04	724 150 792	6502 8741 1216	333704	29·9666481	9·6477368	1113586
899	80 82 01	726 572 699	6531 8885 6401	333519	29·9833287	9·6513166	1112347
900	81 00 00	729 000 000	6561 0000 0000	333333	30·0000000	9·6548938	1111111

No. n	Square n^2	Cube n^3	Fourth power n^4	$\frac{1}{\sqrt{n}}$	Square root $\sqrt{n}$	Cube root $\sqrt[3]{n}$	Reciprocal $\frac{1}{n}$
				0·0			0·00
900	81 00 00	729 000 000	6561 0000 0000	333333	30·0000000	9·6548938	1111111
901	81 18 01	731 432 701	6590 2086 3601	333148	30·0166620	9·6584684	1109878
902	81 36 04	733 870 808	6619 5146 8816	332964	30·0333148	9·6620403	1108647
903	81 54 09	736 314 327	6648 9183 7281	332779	30·0499584	9·6656096	1107420
904	81 72 16	738 763 264	6678 4199 0656	332595	30·0665928	9·6691763	1106195
905	81 90 25	741 217 625	6708 0195 0625	332411	30·0832179	9·6727403	1104972
906	82 08 36	743 677 416	6737 7173 8896	332228	30·0998339	9·6763017	1103753
907	82 26 49	746 142 643	6767 5137 7201	332045	30·1164407	9·6798604	1102536
908	82 44 64	748 613 312	6797 4088 7296	331862	30·1330383	9·6834166	1101322
909	82 62 81	751 089 429	6827 4029 0961	331679	30·1496269	9·6869701	1100110
910	82 81 00	753 571 000	6857 4961 0000	331497	30·1662063	9·6905211	1098901
911	82 99 21	756 058 031	6887 6886 6241	331315	30·1827765	9·6940694	1097695
912	83 17 44	758 550 528	6917 9808 1536	331133	30·1993377	9·6976152	1096491
913	83 35 69	761 048 497	6948 3727 7761	330952	30·2158899	9·7011583	1095290
914	83 53 96	763 551 944	6978 8647 6816	330771	30·2324329	9·7046989	1094092
915	83 72 25	766 060 875	7009 4570 0625	330590	30·2489669	9·7082369	1092896
916	83 90 56	768 575 296	7040 1497 1136	330409	30·2654919	9·7117723	1091703
917	84 08 89	771 095 213	7070 9431 0321	330229	30·2820079	9·7153051	1090513
918	84 27 24	773 620 632	7101 8374 0176	330049	30·2985148	9·7188354	1089325
919	84 45 61	776 151 559	7132 8328 2721	329870	30·3150128	9·7223631	1088139
920	84 64 00	778 688 000	7163 9296 0000	329690	30·3315018	9·7258883	1086957
921	84 82 41	781 229 961	7195 1279 4081	329511	30·3479818	9·7294109	1085776
922	85 00 84	783 777 448	7226 4280 7056	329332	30·3644529	9·7329309	1084599
923	85 19 29	786 330 467	7257 8302 1041	329154	30·3809151	9·7364484	1083424
924	85 37 76	788 889 024	7289 3345 8176	328976	30·3973683	9·7399634	1082251
925	85 56 25	791 453 125	7320 9414 0625	328798	30·4138127	9·7434758	1081081
926	85 74 76	794 022 776	7352 6509 0576	328620	30·4302481	9·7469857	1079914
927	85 93 29	796 597 983	7384 4633 0241	328443	30·4466747	9·7504931	1078749
928	86 11 84	799 178 752	7416 3788 1856	328266	30·4630924	9·7539979	1077586
929	86 30 41	801 765 089	7448 3976 7681	328089	30·4795013	9·7575003	1076426
930	86 49 00	804 357 000	7480 5201 0000	327913	30·4959014	9·7610001	1075269
931	86 67 61	806 954 491	7512 7463 1121	327737	30·5122926	9·7644974	1074114
932	86 86 24	809 557 568	7545 0765 3376	327561	30·5286750	9·7679922	1072961
933	87 04 89	812 166 237	7577 5109 9121	327385	30·5450487	9·7714845	1071811
934	87 23 56	814 780 504	7610 0499 0736	327210	30·5614136	9·7749743	1070664
935	87 42 25	817 400 375	7642 6935 0625	327035	30·5777697	9·7784617	1069519
936	87 60 96	820 025 856	7675 4420 1216	326860	30·5941171	9·7819465	1068376
937	87 79 69	822 656 953	7708 2956 4961	326686	30·6104557	9·7854289	1067236
938	87 98 44	825 293 672	7741 2546 4336	326512	30·6267857	9·7889087	1066098
939	88 17 21	827 936 019	7774 3192 1841	326338	30·6431069	9·7923861	1064963
940	88 36 00	830 584 000	7807 4896 0000	326164	30·6594194	9·7958611	1063830
941	88 54 81	833 237 621	7840 7660 1361	325991	30·6757233	9·7993336	1062699
942	88 73 64	835 896 888	7874 1486 8496	325818	30·6920185	9·8028036	1061571
943	88 92 49	838 561 807	7907 6378 4001	325645	30·7083051	9·8062711	1060445
944	89 11 36	841 232 384	7941 2337 0496	325472	30·7245830	9·8097363	1059322
945	89 30 25	843 908 625	7974 9365 0625	325300	30·7408523	9·8131989	1058201
946	89 49 16	846 590 536	8008 7464 7056	325128	30·7571130	9·8166592	1057082
947	89 68 09	849 278 123	8042 6638 2481	324956	30·7733651	9·8201169	1055966
948	89 87 04	851 971 392	8076 6887 9616	324785	30·7896086	9·8235723	1054852
949	90 06 01	854 670 349	8110 8216 1201	324614	30·8058436	9·8270252	1053741
950	90 25 00	857 375 000	8145 0625 0000	324443	30·8220700	9·8304757	1052632

No. n	Square n^2	Cube n^3	Fourth power n^4	$\frac{1}{\sqrt{n}}$	Square root $\sqrt{n}$	Cube root $\sqrt[3]{n}$	Recipr. $\frac{1}{n}$
				0·0			0·00
950	90 25 00	857 375 000	8145 0625 0000	324443	30·8220700	9·8304757	1052632
951	90 44 01	860 085 351	8179 4116 8801	324272	30·8382879	9·8339238	1051525
952	90 63 04	862 801 408	8213 8694 0416	324102	30·8544972	9·8373695	1050420
953	90 82 09	865 523 177	8248 4358 7681	323932	30·8706981	9·8408127	1049318
954	91 01 16	868 250 664	8283 1113 3456	323762	30·8868904	9·8442536	1048218
955	91 20 25	870 983 875	8317 8960 0625	323592	30·9030743	9·8476920	1047120
956	91 39 36	873 722 816	8352 7901 2096	323423	30·9192497	9·8511280	1046025
957	91 58 49	876 467 493	8387 7939 0801	323254	30·9354166	9·8545617	1044932
958	91 77 64	879 217 912	8422 9075 9696	323085	30·9515751	9·8579929	1043841
959	91 96 81	881 974 079	8458 1314 1761	322917	30·9677251	9·8614218	1042753
960	92 16 00	884 736 000	8493 4656 0000	322749	30·9838668	9·8648483	1041667
961	92 35 21	887 503 681	8528 9103 7441	322581	31·0000000	9·8682724	1040583
962	92 54 44	890 277 128	8564 4659 7136	322413	31·0161248	9·8716941	1039501
963	92 73 69	893 056 347	8600 1326 2161	322245	31·0322413	9·8751135	1038422
964	92 92 96	895 841 344	8635 9105 5616	322078	31·0483494	9·8785305	1037344
965	93 12 25	898 632 125	8671 8000 0625	321911	31·0644491	9·8819451	1036269
966	93 31 56	901 428 696	8707 8012 0336	321745	31·0805405	9·8853574	1035197
967	93 50 89	904 231 063	8743 9143 7921	321578	31·0966236	9·8887673	1034126
968	93 70 24	907 039 232	8780 1397 6576	321412	31·1126984	9·8921749	1033058
969	93 89 61	909 853 209	8816 4775 9521	321246	31·1287648	9·8955801	1031992
970	94 09 00	912 673 000	8852 9281 0000	321081	31·1448230	9·8989830	1030928
971	94 28 41	915 498 611	8889 4915 1281	320915	31·1608729	9·9023835	1029866
972	94 47 84	918 330 048	8926 1680 6656	320750	31·1769145	9·9057817	1028807
973	94 67 29	921 167 317	8962 9579 9441	320585	31·1929479	9·9091776	1027749
974	94 86 76	924 010 424	8999 8615 2976	320421	31·2089731	9·9125712	1026694
975	95 06 25	926 859 375	9036 8789 0625	320256	31·2249900	9·9159624	1025641
976	95 25 76	929 714 176	9074 0103 5776	320092	31·2409987	9·9193513	1024590
977	95 45 29	932 574 833	9111 2561 1841	319928	31·2569992	9·9227379	1023541
978	95 64 84	935 441 352	9148 6164 2256	319765	31·2729915	9·9261222	1022495
979	95 84 41	938 313 739	9186 0915 0481	319601	31·2889757	9·9295042	1021450
980	96 04 00	941 192 000	9223 6816 0000	319438	31·3049517	9·9328839	1020408
981	96 23 61	944 076 141	9261 3869 4321	319275	31·3209195	9·9362613	1019368
982	96 43 24	946 966 168	9299 2077 6976	319113	31·3368792	9·9396364	1018330
983	96 62 89	949 862 087	9337 1443 1521	318950	31·3528308	9·9430092	1017294
984	96 82 56	952 763 904	9375 1968 1536	318788	31·3687743	9·9463797	1016260
985	97 02 25	955 671 625	9413 3655 0625	318626	31·3847097	9·9497479	1015228
986	97 21 96	958 585 256	9451 6506 2416	318465	31·4006369	9·9531138	1014199
987	97 41 69	961 504 803	9490 0524 0561	318304	31·4165561	9·9564775	1013171
988	97 61 44	964 430 272	9528 5710 8736	318142	31·4324673	9·9598389	1012146
989	97 81 21	967 361 669	9567 2069 0641	317982	31·4483704	9·9631981	1011122
990	98 01 00	970 299 000	9605 9601 0000	317821	31·4642654	9·9665549	1010101
991	98 20 81	973 242 271	9644 8309 0561	317660	31·4801525	9·9699095	1009082
992	98 40 64	976 191 488	9683 8195 6096	317500	31·4960315	9·9732619	1008065
993	98 60 49	979 146 657	9722 9263 0401	317340	31·5119025	9·9766120	1007049
994	98 80 36	982 107 784	9762 1513 7296	317181	31·5277655	9·9799599	1006036
995	99 00 25	985 074 875	9801 4950 0625	317021	31·5436206	9·9833055	1005025
996	99 20 16	988 047 936	9840 9574 4256	316862	31·5594677	9·9866489	1004016
997	99 40 09	991 026 973	9880 5389 2081	316703	31·5753068	9·9899900	1003009
998	99 60 04	994 011 992	9920 2396 8016	316544	31·5911380	9·9933289	1002004
999	99 80 01	997 002 999	9960 0599 6001	316386	31·6069613	9·9966656	1001001
1000	1 00 00 00	1 000 000 000	1 0000 0000 0000	316228	31·6227766	10·0000000	1000000

13

Logarithmic Tables

How to use log tables?

Logarithm of a number consists of two parts : (*a*) integral part, called the characteristic and (*b*) decimal part, called the mentissa.

Characteristic of a number between 1 to 9 is zero ;

Characteristic of a number between 10 to 99 is one ;

Characteristic of a number between 100 to 999 is two ;

Remember that, the characteristic of any number is less by one than the number of figures to the left of the decimal point. For example, what is the characteristic of 597.50 ?

See that, there are 3 figures to the left of the decimal point. Thus, the characteristic is 3 − 1 = 2

∴ log of 597.50 = 2 mantissa.

Referring to the tables, find the first two numbers *i.e.*, 59 at the extreme left ; then passing along the horizontal line to the number in the vertical sobemn headed by the third figure 7, we obtain the number 7760. Along the same horizontal line see the mean difference of 5 the fourth number.

Thus, log of 597 = 2.7760

Mean difference for 5 = 4

∴ log of 597.5 = 2.7764

7764 is mantissa

log of 59.75 = 1.7764

log of 5.975 = 0.7764

log of .5975 = $\bar{1}$.7764

log of .05975 = $\bar{2}$.7764

log of .005974 = $\bar{3}$.7764 and so on

Bar 1, ($\bar{1}$) means − 1

Bar ($\bar{2}$) 2, means − 2, etc.

Antilogarithms :

The number corresponding to a given logarithm is found by using the antilogarithm tables.

Let us find the number whose log is 1.9735.

Here, 1 is characteristic and
9735 is mantissa. Now, forget about the characteristic and see the antilog of 973. Then, add to it the mean difference for 5.

Antilog of	973=0.9397
Mean difference for	5= 11
	0.9408

Characteristic is 1

∴ the decimal point will be placed after 2 figures (1+1) from left *i.e.*, number whose log is 1.9735 is 94.08

If the characteristic is 2, then it will be 940.0 *i.e.*, the decimal point will come after figures (2+1) from the left.

Multiplication :

Rule : Add the logarithms of the numbers; the sum is the logarithm of the product ; To get the product, see antilogaritm.

Ex. 69.46×.007358

log of 69.46=1.8418

log of .007358=$\bar{3}$.8668

Sum=$\bar{1}$.7086

Antilog of .7086= .5112

∴ the product is 0.5112

Division :

Rule : Subtract the logarithm of the denominator from the logarithm of the numerator. The difference is the logarithm of the quotient. To get the quotient, see antilogarithm.

.0962÷13.84

log of .0962=$\bar{2}$.9832

log of 13.84=1.1412

Difference=$\bar{3}$.8420

Antilog of .8420=.6950

∴ the quotient is .00695

COMMON LOGARITHMS

	0	1	2	3	4	5	6	7	8	9	Mean Differences								
											1	2	3	4	5	6	7	8	9
10	0000	0043	0086	0128	0170	0212	0253	0294	0334	0374	4	8	12	17	21	25	29	33	37
11	0414	0453	0492	0531	0569	0607	0645	0682	0719	0755	4	8	11	15	19	23	26	30	34
12	0792	0828	0864	0899	0934	0969	1004	1038	1072	1106	3	7	10	14	17	21	24	28	31
13	1139	1173	1206	1239	1271	1303	1335	1367	1399	1430	3	6	10	13	16	19	23	26	29
14	1461	1492	1523	1553	1584	1614	1644	1673	1703	1732	3	6	9	12	15	18	21	24	27
15	1761	1790	1818	1847	1875	1903	1931	1959	1987	2014	3	6	8	11	14	17	20	22	25
16	2041	2068	2095	2122	2148	2175	2201	2227	2253	2279	3	5	8	11	13	16	18	21	24
17	2304	2330	2355	2380	2405	2430	2455	2480	2504	2529	2	5	7	10	12	15	17	20	22
18	2553	2577	2601	2625	2648	2672	2695	2718	2742	2765	2	5	7	9	12	14	16	19	21
19	2788	2810	2833	2856	2878	2900	2923	2945	2967	2989	2	4	7	9	11	13	16	18	20
20	3010	3032	3054	3075	3096	3118	3139	3160	3181	3201	2	4	6	8	11	13	15	17	19
21	3222	3243	3263	3284	3304	3324	3345	3365	3385	3404	2	4	6	8	10	12	14	16	18
22	3424	3444	3464	3483	3502	3522	3541	3560	3579	3598	2	4	6	8	10	12	14	15	17
23	3617	3636	3655	3674	3692	3711	3729	3747	3766	3784	2	4	6	7	9	11	13	15	17
24	3802	3820	3838	3856	3874	3892	3909	3927	3945	3962	2	4	5	7	9	11	12	14	16
25	3979	3997	4014	4031	4048	4065	4082	4099	4116	4133	2	3	5	7	9	10	12	14	15
26	4150	4166	4183	4200	4216	4232	4249	4265	4281	4298	2	3	5	7	8	10	11	13	15
27	4314	4330	4346	4362	4378	4393	4409	4425	4440	4456	2	3	5	6	8	9	11	13	14
28	4472	4487	4502	4518	4533	4548	4564	4579	4594	4609	2	3	5	6	8	9	11	12	14
29	4624	4639	4654	4669	4683	4698	4713	4728	4742	4757	1	3	4	6	7	9	10	12	13
30	4771	4786	4800	4814	4829	4843	4857	4871	4886	4900	1	3	4	6	7	9	10	11	13
31	4914	4928	4942	4955	4969	4983	4997	5011	5024	5038	1	3	4	6	7	8	10	11	12
32	5051	5065	5079	5092	5105	5119	5132	5145	5159	5172	1	3	4	5	7	8	9	11	12
33	5185	5198	5211	5224	5237	5250	5263	5276	5289	5302	1	3	4	5	6	8	9	10	12
34	5315	5328	5340	5353	5366	5378	5391	5403	5416	5428	1	3	4	5	6	8	9	10	11
35	5441	5453	5465	5478	5490	5502	5514	5527	5539	5551	1	2	4	5	6	7	9	10	11
36	5563	5575	5587	5599	5611	5623	5635	5647	5658	5670	1	2	4	5	6	7	8	10	11
37	5682	5694	5705	5717	5729	5740	5752	5763	5775	5786	1	2	3	5	6	7	8	9	10
38	5798	5809	5821	5832	5843	5855	5866	5877	5888	5899	1	2	3	5	6	7	8	9	10
39	5911	5922	5933	5944	5955	5966	5977	5988	5999	6010	1	2	3	4	5	7	8	9	10
40	6021	6031	6042	6053	6064	6075	6085	6096	6107	6117	1	2	3	4	5	6	8	9	10
41	6128	6138	6149	6160	6170	6180	6191	6201	6212	6222	1	2	3	4	5	6	7	8	9
42	6232	6243	6253	6263	6274	6284	6294	6304	6314	6325	1	2	3	4	5	6	7	8	9
43	6335	6345	6355	6365	6375	6385	6395	6405	6415	6425	1	2	3	4	5	6	7	8	9
44	6435	6444	6454	6464	6474	6484	6493	6503	6513	6522	1	2	3	4	5	6	7	8	9
45	6532	6542	6551	6561	6571	6580	6590	6599	6609	6618	1	2	3	4	5	6	7	8	9
46	6628	6637	6646	6656	6665	6675	6684	6693	6702	6712	1	2	3	4	5	6	7	7	8
47	6721	6730	6739	6749	6758	6767	6776	6785	6794	6803	1	2	3	4	5	5	6	7	8
48	6812	6821	6830	6839	6848	6857	6866	6875	6884	6893	1	2	3	4	4	5	6	7	8
49	6902	6911	6920	6928	6937	6946	6955	6964	6972	6981	1	2	3	4	4	5	6	7	8
50	6990	6998	7007	7016	7024	7033	7042	7050	7059	7067	1	2	3	3	4	5	6	7	8
51	7076	7084	7093	7101	7110	7118	7126	7135	7143	7152	1	2	3	3	4	5	6	7	8
52	7160	7168	7177	7185	7193	7202	7210	7218	7226	7235	1	2	2	3	4	5	6	7	7
53	7243	7251	7259	7267	7275	7284	7292	7300	7308	7316	1	2	2	3	4	5	6	6	7
54	7324	7332	7340	7348	7356	7364	7372	7380	7388	7396	1	2	2	3	4	5	6	6	7

COMMON LOGARITHMS

	0	1	2	3	4	5	6	7	8	9	Mean Differences								
											1	2	3	4	5	6	7	8	9
55	7404	7412	7419	7427	7435	7443	7451	7459	7466	7474	1	2	2	3	4	5	5	6	7
56	7482	7490	7497	7505	7513	7520	7528	7536	7543	7551	1	2	2	3	4	5	5	6	7
57	7559	7566	7574	7582	7589	7597	7604	7612	7619	7627	1	2	2	3	4	5	5	6	7
58	7634	7642	7649	7657	7664	7672	7679	7686	7694	7701	1	1	2	3	4	4	5	6	7
59	7709	7716	7723	7731	7738	7745	7752	7760	7767	7774	1	1	2	3	4	4	5	6	7
60	7782	7789	7796	7803	7810	7818	7825	7832	7839	7846	1	1	2	3	4	4	5	6	6
61	7853	7860	7868	7875	7882	7889	7896	7903	7910	7917	1	1	2	3	4	4	5	6	6
62	7924	7931	7938	7945	7952	7959	7966	7973	7980	7987	1	1	2	3	3	4	5	6	6
63	7993	8000	8007	8014	8021	8028	8035	8041	8048	8055	1	1	2	3	3	4	5	5	6
64	8062	8069	8075	8082	8089	8096	8102	8109	8116	8122	1	1	2	3	3	4	5	5	6
65	8129	8136	8142	8149	8156	8162	8169	8176	8182	8189	1	1	2	3	3	4	5	5	6
66	8195	8202	8209	8215	8222	8228	8235	8241	8248	8254	1	1	2	3	3	4	5	5	6
67	8261	8267	8274	8280	8287	8293	8299	8306	8312	8319	1	1	2	3	3	4	5	5	6
68	8325	8331	8338	8344	8351	8357	8363	8370	8376	8382	1	1	2	3	3	4	4	5	6
69	8388	8395	8401	8407	8414	8420	8426	8432	8439	8445	1	1	2	2	3	4	4	5	6
70	8451	8457	8463	8470	8476	8482	8488	8494	8500	8506	1	1	2	2	3	4	4	5	6
71	8513	8519	8525	8531	8537	8543	8549	8555	8561	8567	1	1	2	2	3	4	4	5	5
72	8573	8579	8585	8591	8597	8603	8609	8615	8621	8627	1	1	2	2	3	4	4	5	5
73	8633	8639	8645	8651	8657	8663	8669	8675	8681	8686	1	1	2	2	3	4	4	5	5
74	8692	8698	8704	8710	8716	8722	8727	8733	8739	8745	1	1	2	2	3	4	4	5	5
75	8751	8756	8762	8768	8774	8779	8785	8791	8797	8802	1	1	2	2	3	3	4	5	5
76	8808	8814	8820	8825	8831	8837	8842	8848	8854	8859	1	1	2	2	3	3	4	5	5
77	8865	8871	8876	8882	8887	8893	8899	8904	8910	8915	1	1	2	2	3	3	4	4	5
78	8921	8927	8932	8938	8943	8949	8954	8960	8965	8971	1	1	2	2	3	3	4	4	5
79	8976	8982	8987	8993	8998	9004	9009	9015	9020	9025	1	1	2	2	3	3	4	4	5
80	9031	9036	9042	9047	9053	9058	9063	9069	9074	9079	1	1	2	2	3	3	4	4	5
81	9085	9090	9096	9101	9106	9112	9117	9122	9128	9133	1	1	2	2	3	3	4	4	5
82	9138	9143	9149	9154	9159	9165	9170	9175	9180	9186	1	1	2	2	3	3	4	4	5
83	9191	9196	9201	9206	9212	9217	9222	9227	9232	9238	1	1	2	2	3	3	4	4	5
84	9243	9248	9253	9258	9263	9269	9274	9279	9284	9289	1	1	2	2	3	3	4	4	5
85	9294	9299	9304	9309	9315	9320	9325	9330	9335	9340	1	1	2	2	3	3	4	4	5
86	9345	9350	9355	9360	9365	9370	9375	9380	9385	9390	1	1	2	2	3	3	4	4	5
87	9395	9400	9405	9410	9415	9420	9425	9430	9435	9440	0	1	1	2	2	3	3	4	4
88	9445	9450	9455	9460	9465	9469	9474	9479	9484	9489	0	1	1	2	2	3	3	4	4
89	9494	9499	9504	9509	9513	9518	9523	9528	9533	9538	0	1	1	2	2	3	3	4	4
90	9542	9547	9552	9557	9562	9566	9571	9576	9581	9586	0	1	1	2	2	3	3	4	4
91	9590	9595	9600	9605	9609	9614	9619	9624	9628	9633	0	1	1	2	2	3	3	4	4
92	9638	9643	9647	9652	9657	9661	9666	9671	9675	9680	0	1	1	2	2	3	3	4	4
93	9685	9689	9694	9699	9703	9708	9713	9717	9722	9727	0	1	1	2	2	3	3	4	4
94	9731	9736	9741	9745	9750	9754	9759	9763	9768	9773	0	1	1	2	2	3	3	4	4
95	9777	9782	9786	9791	9795	9800	9805	9809	9814	9818	0	1	1	2	2	3	3	4	4
96	9823	9827	9832	9836	9841	9845	9850	9854	9859	9863	0	1	1	2	2	3	3	4	4
97	9868	9872	9877	9881	9886	9890	9894	9899	9903	9908	0	1	1	2	2	3	3	4	4
98	9912	9917	9921	9926	9930	9934	9939	9943	9948	9952	0	1	1	2	2	3	3	4	4
99	9956	9961	9965	9969	9974	9978	9983	9987	9991	9996	0	1	1	2	2	3	3	3	4

ANTILOGARITHMS

	0	1	2	3	4	5	6	7	8	9	Mean Differences								
											1	2	3	4	5	6	7	8	9
·00	1000	1002	1005	1007	1009	1012	1014	1016	1019	1021	0	0	1	1	1	1	2	2	2
·01	1023	1026	1028	1030	1033	1035	1038	1040	1042	1045	0	0	1	1	1	1	2	2	2
·02	1047	1050	1052	1054	1057	1059	1062	1064	1067	1069	0	0	1	1	1	1	2	2	2
·03	1072	1074	1076	1079	1081	1084	1086	1089	1091	1094	0	0	1	1	1	1	2	2	2
·04	1096	1099	1102	1104	1107	1109	1112	1114	1117	1119	0	1	1	1	1	2	2	2	2
.05	1122	1125	1127	1130	1132	1135	1138	1140	1143	1146	0	1	1	1	1	2	2	2	2
.06	1148	1151	1153	1156	1159	1161	1164	1167	1169	1172	0	1	1	1	1	2	2	2	2
.07	1175	1178	1180	1183	1186	1189	1191	1194	1197	1199	0	1	1	1	1	2	2	2	2
.08	1202	1205	1208	1211	1213	1216	1219	1222	1225	1227	0	1	1	1	1	2	2	2	3
·09	1230	1233	1236	1239	1242	1245	1247	1250	1253	1256	0	1	1	1	1	2	2	2	3
·10	1259	1262	1265	1268	1271	1274	1276	1279	1282	1285	0	1	1	1	1	2	2	2	3
·11	1288	1291	1294	1297	1300	1303	1306	1309	1312	1315	0	1	1	1	2	2	2	2	3
·12	1318	1321	1324	1327	1330	1334	1337	1340	1343	1346	0	1	1	1	2	2	2	2	3
·13	1349	1352	1355	1358	1361	1365	1368	1371	1374	1377	0	1	1	1	2	2	2	3	3
·14	1380	1384	1387	1390	1393	1396	1400	1403	1406	1409	0	1	1	1	2	2	2	3	3
·15	1413	1416	1419	1422	1426	1429	1432	1435	1439	1442	0	1	1	1	2	2	2	3	3
·16	1445	1449	1452	1455	1459	1462	1466	1469	1472	1476	0	1	1	1	2	2	2	3	3
·17	1479	1483	1486	1489	1493	1496	1500	1503	1507	1510	0	1	1	1	2	2	2	3	3
·18	1514	1517	1521	1524	1528	1531	1535	1538	1542	1545	0	1	1	1	2	2	2	3	3
·19	1549	1552	1556	1560	1563	1567	1570	1574	1578	1581	0	1	1	1	2	2	3	3	3
·20	1585	1589	1592	1596	1600	1603	1607	1611	1614	1618	0	1	1	1	2	2	3	3	3
·21	1622	1626	1629	1633	1637	1641	1644	1648	1652	1656	0	1	1	2	2	2	3	3	3
·22	1660	1663	1667	1671	1675	1679	1683	1687	1690	1694	0	1	1	2	2	2	3	3	3
·23	1698	1702	1706	1710	1714	1718	1722	1726	1730	1734	0	1	1	2	2	2	3	3	4
·24	1738	1742	1746	1750	1754	1758	1762	1766	1770	1774	0	1	1	2	2	2	3	3	4
·25	1778	1782	1786	1791	1795	1799	1803	1807	1811	1816	0	1	1	2	2	2	3	3	4
·26	1820	1824	1828	1832	1837	1841	1845	1849	1854	1858	0	1	1	2	2	3	3	3	4
·27	1862	1866	1871	1875	1879	1884	1888	1892	1897	1901	0	1	1	2	2	3	3	3	4
·28	1905	1910	1914	1919	1923	1928	1932	1936	1941	1945	0	1	1	2	2	3	3	4	4
·29	1950	1954	1959	1963	1968	1972	1977	1982	1986	1991	0	1	1	2	2	3	3	4	4
·30	1995	2000	2004	2009	2014	2018	2023	2028	2032	2037	0	1	1	2	2	3	3	4	4
·31	2042	2046	2051	2056	2061	2065	2070	2075	2080	2084	0	1	1	2	2	3	3	4	4
·32	2089	2094	2099	2104	2109	2113	2118	2123	2128	2133	0	1	1	2	2	3	3	4	4
·33	2138	2143	2148	2153	2158	2163	2168	2173	2178	2183	0	1	1	2	2	3	3	4	4
·34	2188	2193	2198	2203	2208	2213	2218	2223	2228	2234	1	1	2	2	3	3	4	4	5
·35	2239	2244	2249	2254	2259	2265	2270	2275	2280	2286	1	1	2	2	3	3	4	4	5
·36	2291	2296	2301	2307	2312	2317	2323	2328	2333	2339	1	1	2	2	3	3	4	4	5
·37	2344	2350	2355	2360	2366	2371	2377	2382	2388	2393	1	1	2	2	3	3	4	4	5
·38	2399	2404	2410	2415	2421	2427	2432	2438	2443	2449	1	1	2	2	3	3	4	4	5
·39	2455	2460	2466	2472	2477	2483	2489	2495	2500	2506	1	1	2	2	3	3	4	5	5
·40	2512	2518	2523	2529	2535	2541	2547	2553	2559	2564	1	1	2	2	3	4	4	5	5
·41	2570	2576	2582	2588	2594	2600	2606	2612	2618	2624	1	1	2	2	3	4	4	5	5
·42	2630	2636	2642	2649	2655	2661	2667	2673	2679	2685	1	1	2	2	3	4	4	5	6
·43	2692	2698	2704	2710	2716	2723	2729	2735	2742	2748	1	1	2	3	3	4	4	5	6
·44	2754	2761	2767	2773	2780	2786	2793	2799	2805	2812	1	1	2	3	3	4	4	5	6
·45	2818	2825	2831	2838	2844	2851	2858	2864	2871	2877	1	1	2	3	3	4	5	5	6
·46	2884	2891	2897	2904	2911	2917	2924	2931	2938	2944	1	1	2	3	3	4	5	5	6
·47	2951	2958	2965	2972	2979	2985	2992	2999	3006	3013	1	1	2	3	3	4	5	5	6
·48	3020	3027	3034	3041	3048	3055	3062	3069	3076	3083	1	1	2	3	4	4	5	6	6
·49	3090	3097	3105	3112	3119	3126	3133	3141	3148	3155	1	1	2	3	4	4	5	6	6

ANTILOGARITHMS

	0	1	2	3	4	5	6	7	8	9	Mean Differences								
											1	2	3	4	5	6	7	8	9
·50	3162	3170	3177	3184	3192	3199	3206	3214	3221	3228	1	1	2	3	4	4	5	6	7
·51	3236	3243	3251	3258	3266	3273	3281	3289	3296	3304	1	2	2	3	4	5	5	6	7
·52	3311	3319	3327	3334	3342	3350	3357	3365	3373	3381	1	2	2	3	4	5	5	6	7
·53	3388	3396	3404	3412	3420	3428	3436	3443	3451	3459	1	2	2	3	4	5	6	6	7
·54	3467	3475	3483	3491	3499	3508	3516	3524	3532	3540	1	2	2	3	4	5	6	6	7
·55	3548	3556	3565	3573	3581	3589	3597	3606	3614	3622	1	2	2	3	4	5	6	7	7
·56	3631	3639	3648	3656	3664	3673	3681	3690	3698	3707	1	2	3	3	4	5	6	7	8
·57	3715	3724	3733	3741	3750	3758	3767	3776	3784	3793	1	2	3	3	4	5	6	7	8
·58	3802	3811	3819	3828	3837	3846	3855	3864	3873	3882	1	2	3	4	4	5	6	7	8
·59	3890	3899	3908	3917	3926	3936	3945	3954	3963	3972	1	2	3	4	5	5	6	7	8
·60	3981	3990	3999	4009	4018	4027	4036	4046	4055	4064	1	2	3	4	5	6	6	7	8
·61	4074	4083	4093	4102	4111	4121	4130	4140	4150	4159	1	2	3	4	5	6	7	8	9
·62	4169	4178	4188	4198	4207	4217	4227	4236	4246	4256	1	2	3	4	5	6	7	8	9
·63	4266	4276	4285	4295	4305	4315	4325	4335	4345	4355	1	2	3	4	5	6	7	8	9
·64	4365	4375	4385	4395	4406	4416	4426	4436	4446	4457	1	2	3	4	5	6	7	8	9
·65	4467	4477	4487	4498	4508	4519	4529	4539	4550	4560	1	2	3	4	5	6	7	8	9
·66	4571	4581	4592	4603	4613	4624	4634	4645	4656	4667	1	2	3	4	5	6	7	9	10
·67	4677	4688	4699	4710	4721	4732	4742	4753	4764	4775	1	2	3	4	5	7	8	9	10
·68	4786	4797	4808	4819	4831	4842	4853	4864	4875	4887	1	2	3	4	6	7	8	9	10
·69	4898	4909	4920	4932	4943	4955	4966	4977	4989	5000	1	2	3	5	6	7	8	9	10
·70	5012	5023	5035	5047	5058	5070	5082	5093	5105	5117	1	2	4	5	6	7	8	9	11
·71	5129	5140	5152	5164	5176	5188	5200	5212	5224	5236	1	2	4	5	6	7	8	10	11
·72	5248	5260	5272	5284	5297	5309	5321	5333	5346	5358	1	2	4	5	6	7	9	10	11
·73	5370	5383	5395	5408	5420	5433	5445	5458	5470	5483	1	3	4	5	6	8	9	10	11
·74	5495	5508	5521	5534	5546	5559	5572	5585	5598	5610	1	3	4	5	6	8	9	10	12
·75	5623	5636	5649	5662	5675	5689	5702	5715	5728	5741	1	3	4	5	7	8	9	10	12
·76	5754	5768	5781	5794	5808	5821	5834	5848	5861	5875	1	3	4	5	7	8	9	11	12
·77	5888	5902	5916	5929	5943	5957	5970	5984	5998	6012	1	3	4	5	7	8	10	11	12
·78	6026	6039	6053	6067	6081	6095	6109	6124	6138	6152	1	3	4	6	7	8	10	11	13
·79	6166	6180	6194	6209	6223	6237	6252	6266	6281	6295	1	3	4	6	7	9	10	11	13
·80	6310	6324	6339	6353	6368	6383	6397	6412	6427	6442	1	3	4	6	7	9	10	12	13
·81	6457	6471	6486	6501	6516	6531	6546	6561	6577	6592	2	3	5	6	8	9	11	12	14
·82	6607	6622	6637	6653	6668	6683	6699	6714	6730	6745	2	3	5	6	8	9	11	12	14
·83	6761	6776	6792	6808	6823	6839	6855	6871	6887	6902	2	3	5	6	8	9	11	13	14
·84	6918	6934	6950	6966	6982	6998	7015	7031	7047	7063	2	3	5	6	8	10	11	13	15
·85	7079	7096	7112	7129	7145	7161	7178	7194	7211	7228	2	3	5	7	8	10	12	13	15
·86	7244	7261	7278	7295	7311	7328	7345	7362	7379	7396	2	3	5	7	8	10	12	13	15
·87	7413	7430	7447	7464	7482	7499	7516	7534	7551	7568	2	3	5	7	9	10	12	14	16
·88	7586	7603	7621	7638	7656	7674	7691	7709	7727	7745	2	4	5	7	9	11	12	14	16
·89	7762	7780	7798	7816	7834	7852	7870	7889	7907	7925	2	4	5	7	9	11	13	14	16
·90	7943	7962	7980	7998	8017	8035	8054	8072	8091	8110	2	4	6	7	9	11	13	15	17
·91	8128	8147	8166	8185	8204	8222	8241	8260	8279	8299	2	4	6	8	9	11	13	15	17
·92	8318	8337	8356	8375	8395	8414	8433	8453	8472	8492	2	4	6	8	10	12	14	15	17
·93	8511	8531	8551	8570	8590	8610	8630	8650	8670	8690	2	4	6	8	10	12	14	16	18
·94	8710	8730	8750	8770	8790	8810	8831	8851	8872	8892	2	4	6	8	10	12	14	16	18
·95	8913	8933	8954	8974	8995	9016	9036	9057	9078	9099	2	4	6	8	10	12	15	17	19
·96	9120	9141	9162	9183	9204	9226	9247	9268	9290	9311	2	4	6	8	11	13	15	17	19
·97	9333	9354	9376	9397	9419	9441	9462	9484	9506	9528	2	4	7	9	11	13	15	17	20
·98	9550	9572	9594	9616	9638	9661	9683	9705	9727	9750	2	4	7	9	11	13	16	18	20
·99	9772	9795	9817	9840	9863	9886	9908	9931	9954	9977	2	5	7	9	11	14	16	18	20

NATURAL SINES

Degree	0′	6′	12′	18′	24′	30′	36′	42′	48′	54′	Mean Differences 1′	2′	3′	4′	5′
0	·0000	0017	0035	0052	0070	0087	0105	0122	0140	0157	3	6	9	12	15
1	·0175	0192	0209	0227	0244	0262	0279	0297	0314	0332	3	6	9	12	15
2	·0349	0366	0384	0401	0419	0436	0454	0471	0488	0506	3	6	9	12	15
3	·0523	0541	0558	0576	0593	0610	0628	0645	0663	0680	3	6	9	12	15
4	·0698	0715	0732	0750	0767	0785	0802	0819	0837	0854	3	6	9	12	15
5	·0872	0889	0906	0924	0941	0958	0976	0993	1011	1028	3	6	9	12	14
6	·1045	1063	1080	1097	1115	1132	1149	1167	1184	1201	3	6	9	12	14
7	·1219	1236	1253	1271	1288	1305	1323	1340	1357	1374	3	6	9	12	14
8	·1392	1409	1426	1444	1461	1478	1495	1513	1530	1547	3	6	9	12	14
9	·1564	1582	1599	1616	1633	1650	1668	1685	1702	1719	3	6	9	12	14
10	·1736	1754	1771	1788	1805	1822	1840	1857	1874	1891	3	6	9	12	14
11	·1908	1925	1942	1959	1977	1994	2011	2028	2045	2062	3	6	9	11	14
12	·2079	2096	2113	2130	2147	2164	2181	2198	2215	2232	3	6	9	11	14
13	·2250	2267	2284	2300	2317	2334	2351	2368	2385	2402	3	6	8	11	14
14	·2419	2436	2453	2470	2487	2504	2521	2538	2554	2571	3	6	8	11	14
15	·2588	2605	2622	2639	2656	2672	2689	2706	2723	2740	3	6	8	11	14
16	·2756	2773	2790	2807	2823	2840	2857	2874	2890	2907	3	6	8	11	14
17	·2924	2940	2957	2974	2990	3007	3024	3040	3057	3074	3	6	8	11	14
18	·3090	3107	3123	3140	3156	3173	3190	3206	3223	3239	3	6	8	11	14
19	·3256	3272	3289	3305	3322	3338	3355	3371	3387	3404	3	5	8	11	14
20	·3420	3437	3453	3469	3486	3502	3518	3535	3551	3567	3	5	8	11	14
21	·3584	3600	3616	3633	3649	3665	3681	3697	3714	3730	3	5	8	11	14
22	·3746	3762	3778	3795	3811	3827	3843	3859	3875	3891	3	5	8	11	14
23	·3907	3923	3939	3955	3971	3987	4003	4019	4035	4051	3	5	8	11	14
24	·4067	4083	4099	4115	4131	4147	4163	4179	4195	4210	3	5	8	11	13
25	·4226	4242	4258	4274	4289	4305	4321	4337	4352	4368	3	5	8	11	13
26	·4384	4399	4415	4431	4446	4462	4478	4493	4509	4524	3	5	8	10	13
27	·4540	4555	4571	4586	4602	4617	4633	4648	4664	4679	3	5	8	10	13
28	·4695	4710	4726	4741	4756	4772	4787	4802	4818	4833	3	5	8	10	13
29	·4848	4863	4879	4894	4909	4924	4939	4955	4970	4985	3	5	8	10	13
30	·5000	5015	5030	5045	5060	5075	5090	5105	5120	5135	3	5	8	10	13
31	·5150	5165	5180	5195	5210	5225	5240	5255	5270	5284	2	5	7	10	12
32	·5299	5314	5329	5344	5358	5373	5388	5402	5417	5432	2	5	7	10	12
33	·5446	5461	5476	5490	5505	5519	5534	5548	5563	5577	2	5	7	10	12
34	·5592	5606	5621	5635	5650	5664	5678	5693	5707	5721	2	5	7	10	12
35	·5736	5750	5764	5779	5793	5807	5821	5835	5850	5864	2	5	7	9	12
36	·5878	5892	5906	5920	5934	5948	5962	5976	5990	6004	2	5	7	9	12
37	·6018	6032	6046	6060	6074	6088	6101	6115	6129	6143	2	5	7	9	12
38	·6157	6170	6184	6198	6211	6225	6239	6252	6266	6280	2	5	7	9	11
39	·6293	6307	6320	6334	6347	6361	6374	6388	6401	6414	2	4	7	9	11
40	·6428	6441	6455	6468	6481	6494	6508	6521	6534	6547	2	4	7	9	11
41	·6561	6574	6587	6600	6613	6626	6639	6652	6665	6678	2	4	7	9	11
42	·6691	6704	6717	6730	6743	6756	6769	6782	6794	6807	2	4	6	9	11
43	·6820	6833	6845	6858	6871	6884	6896	6909	6921	6934	2	4	6	8	11
44	·6947	6959	6972	6984	6997	7009	7022	7034	7046	7059	2	4	6	8	10

NATURAL SINES

Degree	0′	6′	12′	18′	24′	30′	36′	42′	48′	54′	Mean Differences 1′	2′	3′	4′	5′
45	·7071	7083	7096	7108	7120	7133	7145	7157	7169	7181	2	4	6	8	10
46	·7193	7206	7218	7230	7242	7254	7266	7278	7290	7302	2	4	6	8	10
47	·7314	7325	7337	7349	7361	7373	7385	7396	7408	7420	2	4	6	8	10
48	·7431	7443	7455	7466	7478	7490	7501	7513	7524	7536	2	4	6	8	10
49	·7547	7559	7570	7581	7593	7604	7615	7627	7638	7649	2	4	6	8	9
50	·7660	7672	7683	7694	7705	7716	7727	7738	7749	7760	2	4	6	7	9
51	·7771	7782	7793	7804	7815	7826	7837	7848	7859	7869	2	4	5	7	9
52	·7880	7891	7902	7912	7923	7934	7944	7955	7965	7976	2	4	5	7	9
53	·7986	7997	8007	8018	8028	8039	8049	8059	8070	8080	2	3	5	7	9
54	·8090	8100	8111	8121	8131	8141	8151	8161	8171	8181	2	3	5	7	8
55	·8192	8202	8211	8221	8231	8241	8251	8261	8271	8281	2	3	5	7	8
56	·8290	8300	8310	8320	8329	8339	8348	8358	8368	8377	2	3	5	6	8
57	·8387	8396	8406	8415	8425	8434	8443	8453	8462	8471	2	3	5	6	8
58	·8480	8490	8499	8508	8517	8526	8536	8545	8554	8563	2	3	5	6	8
59	·8572	8581	8590	8599	8607	8616	8625	8634	8643	8652	1	3	4	6	7
60	·8660	8669	8678	8686	8695	8704	8712	8721	8729	8738	1	3	4	6	7
61	·8746	8755	8763	8771	8780	8788	8796	8805	8813	8821	1	3	4	6	7
62	·8829	8838	8846	8854	8862	8870	8878	8886	8894	8902	1	3	4	5	7
63	·8910	8918	8926	8934	8942	8949	8957	8965	8973	8980	1	3	4	5	6
64	·8988	8996	9003	9011	9018	9026	9033	9041	9048	9056	1	3	4	5	6
65	·9063	9070	9078	9085	9092	9100	9107	9114	9121	9128	1	2	4	5	6
66	·9135	9143	9150	9157	9164	9171	9178	9184	9191	9198	1	2	3	5	6
67	·9205	9212	9219	9225	9232	9239	9245	9252	9259	9265	1	2	3	4	6
68	·9272	9278	9285	9291	9298	9304	9311	9317	9323	9330	1	2	3	4	5
69	·9336	9342	9348	9354	9361	9367	9373	9379	9385	9391	1	2	3	4	5
70	·9397	9403	9409	9415	9421	9426	9432	9438	9444	9449	1	2	3	4	5
71	·9455	9461	9466	9472	9478	9483	9489	9494	9500	9505	1	2	3	4	5
72	·9511	9516	9521	9527	9532	9537	9542	9548	9553	9558	1	2	3	3	4
73	·9563	9568	9573	9578	9583	9588	9593	9598	9603	9608	1	2	2	3	4
74	·9613	9617	9622	9627	9632	9636	9641	9646	9650	9655	1	2	2	3	4
75	·9659	9664	9668	9673	9677	9681	9686	9690	9694	9699	1	1	2	3	4
76	·9703	9707	9711	9715	9720	9724	9728	9732	9736	9740	1	1	2	3	3
77	·9744	9748	9751	9755	9759	9763	9767	9770	9774	9778	1	1	2	3	3
78	·9781	9785	9789	9792	9796	9799	9803	9806	9810	9813	1	1	2	2	3
79	·9816	9820	9823	9826	9829	9833	9836	9839	9842	9845	1	1	2	2	3
80	·9848	9851	9854	9857	9860	9863	9866	9869	9871	9874	0	1	1	2	2
81	·9877	9880	9882	9885	9888	9890	9893	9895	9898	9900	0	1	1	2	2
82	·9903	9905	9907	9910	9912	9914	9917	9919	9921	9923	0	1	1	2	2
83	·9925	9928	9930	9932	9934	9936	9938	9940	9942	9943	0	1	1	1	2
84	·9945	9947	9949	9951	9952	9954	9956	9957	9959	9960	0	1	1	1	2
85	·9962	9963	9965	9966	9968	9969	9971	9972	9973	9974	0	0	1	1	1
86	·9976	9977	9978	9979	9980	9981	9982	9983	9984	9985	0	0	1	1	1
87	·9986	9987	9988	9989	9990	9990	9991	9992	9993	9993	0	0	0	1	1
88	·9994	9995	9995	9996	9996	9997	9997	9997	9998	9998	0	0	0	0	0
89	·9998	9999	9999	9999	9999	1·000	1·000	1·000	1·000	1·000	0	0	0	0	0

NATURAL COSINES

Degree	0′	6′	12′	18′	24′	30′	36′	42′	48′	54′	Mean Differences 1′	2′	3′	4′	5′
0	1·000	1·000	1·000	1·000	1·000	1·000	·9999	9999	9999	9999	0	0	0	0	0
1	·9998	9998	9998	9997	9997	9997	9996	9996	9995	9995	0	0	0	0	0
2	·9994	9993	9993	9992	9991	9990	9990	9989	9988	9987	0	0	0	1	1
3	·9986	9985	9984	9983	9982	9981	9980	9979	9978	9977	0	0	1	1	1
4	·9976	9974	9973	9972	9971	9969	9968	9966	9965	9963	0	0	1	1	1
5	·9962	9960	9959	9957	9956	9954	9952	9951	9949	9947	0	1	1	1	1
6	·9945	9943	9942	9940	9938	9936	9934	9932	9930	9928	0	1	1	1	2
7	·9925	9923	9921	9919	9917	9914	9912	9910	9907	9905	0	1	1	2	2
8	·9903	9900	9898	9895	9893	9890	9888	9885	9882	9880	0	1	1	2	2
9	·9877	9874	9871	9869	9866	9863	9860	9857	9854	9851	0	1	1	2	2
10	·9848	9845	9842	9839	9836	9833	9829	9826	9823	9820	1	1	2	2	3
11	·9816	9813	9810	9806	9803	9799	9796	9792	9789	9785	1	1	2	2	3
12	·9781	9778	9774	9770	9767	9763	9759	9755	9751	9748	1	1	2	3	3
13	·9744	9740	9736	9732	9728	9724	9720	9715	9711	9707	1	1	2	3	3
14	·9703	9699	9694	9690	9686	9681	9677	9673	9668	9664	1	1	2	3	4
15	·9659	9655	9650	9646	9641	9636	9632	9627	9622	9617	1	2	2	3	4
16	·9613	9608	9603	9598	9593	9588	9583	9578	9573	9568	1	2	2	3	4
17	·9563	9558	9553	9548	9542	9537	9532	9527	9521	9516	1	2	3	3	4
18	·9511	9505	9500	9494	9489	9483	9478	9472	9466	9461	1	2	3	4	5
19	·9455	9449	9444	9438	9432	9426	9421	9415	9409	9403	1	2	3	4	5
20	·9397	9391	9385	9379	9373	9367	9361	9354	9348	9342	1	2	3	4	5
21	·9336	9330	9323	9317	9311	9304	9298	9291	9285	9278	1	2	3	4	5
22	·9272	9265	9259	9252	9245	9239	9232	9225	9219	9212	1	2	3	4	6
23	·9205	9198	9191	9184	9178	9171	9164	9157	9150	9143	1	2	3	5	6
24	·9135	9128	9121	9114	9107	9100	9092	9085	9078	9070	1	2	4	5	6
25	·9063	9056	9048	9041	9033	9026	9018	9011	9003	8996	1	3	4	5	6
26	·8988	8980	8973	8965	8957	8949	8942	8934	8926	8918	1	3	4	5	6
27	·8910	8902	8894	8886	8878	8870	8862	8854	8846	8838	1	3	4	5	7
28	·8829	8821	8813	8805	8796	8788	8780	8771	8763	8755	1	3	4	6	7
29	·8746	8738	8729	8721	8712	8704	8695	8686	8678	8669	1	3	4	6	7
30	·8660	8652	8643	8634	8625	8616	8607	8599	8590	8581	1	3	4	6	7
31	·8572	8563	8554	8545	8536	8526	8517	8508	8499	8490	2	3	5	6	8
32	·8480	8471	8462	8453	8443	8434	8425	8415	8406	8396	2	3	5	6	8
33	·8387	8377	8368	8358	8348	8339	8329	8320	8310	8300	2	3	5	6	8
34	·8290	8281	8271	8261	8251	8241	8231	8221	8211	8202	2	3	5	7	8
35	·8192	8181	8171	8161	8151	8141	8131	8121	8111	8100	2	3	5	7	8
36	·8090	8080	8070	8059	8049	8039	8028	8018	8007	7997	2	3	5	7	9
37	·7986	7976	7965	7955	7944	7934	7923	7912	7902	7891	2	4	5	7	9
38	·7880	7869	7859	7848	7837	7826	7815	7804	7793	7782	2	4	5	7	9
39	·7771	7760	7749	7738	7727	7716	7705	7694	7683	7672	2	4	6	7	9
40	·7660	7649	7638	7627	7615	7604	7593	7581	7570	7559	2	4	6	8	9
41	·7547	7536	7524	7513	7501	7490	7478	7466	7455	7443	2	4	6	8	10
42	·7431	7420	7408	7396	7385	7373	7361	7349	7337	7325	2	4	6	8	10
43	·7314	7302	7290	7278	7266	7254	7242	7230	7218	7206	2	4	6	8	10
44	·7193	7181	7169	7157	7145	7133	7120	7108	7096	7083	2	4	6	8	10

**Subtract mean differences*

NATURAL COSINES

Degree		M	I	N	U	T	E	S			Mean Differences			
	0′	6′	12′	18′	24′	30′	36′	42′	48′	54′	1′	2′	3′	4′	5′
45	·7071	7059	7046	7034	7022	7009	6997	6984	6972	6959	2	4	6	8	10
46	·6947	6934	6921	6909	6896	6884	6871	6858	6845	6833	2	4	6	8	11
47	·6820	6807	6794	6782	6769	6756	6743	6730	6717	6704	2	4	6	9	11
48	·6691	6678	6665	6652	6639	6626	6613	6600	6587	6574	2	4	7	9	11
49	·6561	6547	6534	6521	6508	6494	6481	6468	6455	6441	2	4	7	9	11
50	·6428	6414	6401	6388	6374	6361	6347	6334	6320	6307	2	4	7	9	11
51	·6293	6280	6266	6252	6239	6225	6211	6198	6184	6170	2	5	7	9	11
52	·6157	6143	6129	6115	6101	6088	6074	6060	6046	6032	2	5	7	9	12
53	·6018	6004	5990	5976	5962	5948	5934	5920	5906	5892	2	5	7	9	12
54	·5878	5864	5850	5835	5821	5807	5793	5779	5764	5750	2	5	7	9	12
55	·5736	5721	5707	5693	5678	5664	5650	5635	5621	5606	2	5	7	10	12
56	·5592	5577	5563	5548	5534	5519	5505	5490	5476	5461	2	5	7	10	12
57	·5446	5432	5417	5402	5388	5373	5358	5344	5329	5314	2	5	7	10	12
58	·5299	5284	5270	5255	5240	5225	5210	5195	5180	5165	2	5	7	10	12
59	·5150	5135	5120	5105	5090	5075	5060	5045	5030	5015	3	5	8	10	13
60	·5000	4985	4970	4955	4939	4924	4909	4894	4879	4863	3	5	8	10	13
61	·4848	4833	4818	4802	4787	4772	4756	4741	4726	4710	3	5	8	10	13
62	·4695	4679	4664	4648	4633	4617	4602	4586	4571	4555	3	5	8	10	13
63	·4540	4524	4509	4493	4478	4462	4446	4431	4415	4399	3	5	8	10	13
64	·4384	4368	4352	4337	4321	4305	4289	4274	4258	4242	3	5	8	11	13
65	·4226	4210	4195	4179	4163	4147	4131	4115	4099	4083	3	5	8	11	13
66	·4067	4051	4035	4019	4003	3987	3971	3955	3939	3923	3	5	8	11	14
67	·3907	3891	3875	3859	3843	3827	3811	3795	3778	3762	3	5	8	11	14
68	·3746	3730	3714	3697	3681	3665	3649	3633	3616	3600	3	5	8	11	14
69	·3584	3567	3551	3535	3518	3502	3486	3469	3453	3437	3	5	8	11	14
70	·3420	3404	3387	3371	3355	3338	3322	3305	3289	3272	3	5	8	11	14
71	·3256	3239	3223	3206	3190	3173	3156	3140	3123	3107	3	6	8	11	14
72	·3090	3074	3057	3040	3024	3007	2990	2974	2957	2940	3	6	8	11	14
73	·2924	2907	2890	2874	2857	2840	2823	2807	2790	2773	3	6	8	11	14
74	·2756	2740	2723	2706	2689	2672	2656	2639	2622	2605	3	6	8	11	14
75	·2588	2571	2554	2538	2521	2504	2487	2470	2453	2436	3	6	8	11	14
76	·2419	2402	2385	2368	2351	2334	2317	2300	2284	2267	3	6	8	11	14
77	·2250	2233	2215	2198	2181	2164	2147	2130	2113	2096	3	6	9	11	14
78	·2079	2062	2045	2028	2011	1994	1977	1959	1942	1925	3	6	9	11	14
79	·1908	1891	1874	1857	1840	1822	1805	1788	1771	1754	3	6	9	11	14
80	·1736	1719	1702	1685	1668	1650	1633	1616	1599	1582	3	6	9	12	14
81	·1564	1547	1530	1513	1495	1478	1461	1444	1426	1409	3	6	9	12	14
82	·1392	1374	1357	1340	1323	1305	1288	1271	1253	1236	3	6	9	12	14
83	·1219	1201	1184	1167	1149	1132	1115	1097	1080	1063	3	6	9	12	14
84	·1045	1028	1011	0993	0976	0958	0941	0924	0906	0889	3	6	9	12	14
85	·0872	0854	0837	0819	0802	0785	0767	0750	0732	0715	3	6	9	12	14
86	·0698	0680	0663	0645	0628	0610	0593	0576	0558	0541	3	6	9	12	15
87	·0523	0506	0488	0471	0454	0436	0419	0401	0384	0366	3	6	9	12	15
88	·0349	0332	0314	0297	0279	0262	0244	0227	0209	0192	3	6	9	12	15
89	·0175	0157	0140	0122	0105	0087	0070	0052	0035	0017	3	6	9	12	15

**Subtract mean differences*

NATURAL TANGENTS

Degree	0′	6′	12′ (M)	18′ (I)	24′ (N)	30′ (U)	36′ (T)	42′ (E)	48′ (S)	54′	Mean Differences 1′	2′	3′	4′	5′
0	·0000	0017	0035	0052	0070	0087	0105	0122	0140	0157	3	6	9	12	15
1	·0175	0192	0209	0227	0244	0262	0279	0297	0314	0332	3	6	9	12	15
2	·0349	0367	0384	0402	0419	0437	0454	0472	0489	0507	3	6	9	12	15
3	·0524	0542	0559	0577	0594	0612	0629	0647	0664	0682	3	6	9	12	15
4	·0699	0717	0734	0752	0769	0787	0805	0822	0840	0857	3	6	9	12	15
5	·0875	0892	0910	0928	0945	0963	0981	0998	1016	1033	3	6	9	12	15
6	·1051	1069	1086	1104	1122	1139	1157	1175	1192	1210	3	6	9	12	15
7	·1228	1246	1263	1281	1299	1317	1334	1352	1370	1388	3	6	9	12	15
8	·1405	1423	1441	1459	1477	1495	1512	1530	1548	1566	3	6	9	12	15
9	·1584	1602	1620	1638	1655	1673	1691	1709	1727	1745	3	6	9	12	15
10	·1763	1781	1799	1817	1835	1853	1871	1890	1908	1926	3	6	9	12	15
11	·1944	1962	1980	1998	2016	2035	2053	2071	2089	2107	3	6	9	12	15
12	·2126	2144	2162	2180	2199	2217	2235	2254	2272	2290	3	6	9	12	15
13	·2309	2327	2345	2364	2382	2401	2419	2438	2456	2475	3	6	9	12	15
14	·2493	2512	2530	2549	2568	2586	2605	2623	2642	2661	3	6	9	12	16
15	·2679	2698	2717	2736	2754	2773	2792	2811	2830	2849	3	6	9	13	16
16	·2867	2886	2905	2924	2943	2962	2981	3000	3019	3038	3	6	9	13	16
17	·3057	3076	3096	3115	3134	3153	3172	3191	3211	3230	3	6	10	13	16
18	·3249	3269	3288	3307	3327	3346	3365	3385	3404	3424	3	6	10	13	16
19	·3443	3463	3482	3502	3522	3541	3561	3581	3600	3620	3	7	10	13	16
20	·3640	3659	3679	3699	3719	3739	3759	3779	3799	3819	3	7	10	13	17
21	·3839	3859	3879	3899	3919	3939	3959	3979	4000	4020	3	7	10	13	17
22	·4040	4061	4081	4101	4122	4142	4163	4183	4204	4224	3	7	10	14	17
23	·4245	4265	4286	4307	4327	4348	4369	4390	4411	4431	3	7	10	14	17
24	·4452	4473	4494	4515	4536	4557	4578	4599	4621	4642	4	7	11	14	18
25	·4663	4684	4706	4727	4748	4770	4791	4813	4834	4856	4	7	11	14	18
26	·4877	4899	4921	4942	4964	4986	5008	5029	5051	5073	4	7	11	15	18
27	·5095	5117	5139	5161	5184	5206	5228	5250	5272	5295	4	7	11	15	18
28	·5317	5340	5362	5384	5407	5430	5452	5475	5498	5520	4	8	11	15	19
29	·5543	5566	5589	5612	5635	5658	5681	5704	5727	5750	4	8	12	15	19
30	·5774	5797	5820	5844	5867	5890	5914	5938	5961	5985	4	8	12	16	20
31	·6009	6032	6056	6080	6104	6128	6152	6176	6200	6224	4	8	12	16	20
32	·6249	6273	6297	6322	6346	6371	6395	6420	6445	6469	4	8	12	16	20
33	·6494	6519	6544	6569	6594	6619	6644	6669	6694	6720	4	8	13	17	21
34	·6745	6771	6796	6822	6847	6873	6899	6924	6950	6976	4	9	13	17	21
35	·7002	7028	7054	7080	7107	7133	7159	7186	7212	7239	4	9	13	18	22
36	·7265	7292	7319	7346	7373	7400	7427	7454	7481	7508	5	9	14	18	23
37	·7536	7563	7590	7618	7646	7673	7701	7729	7757	7785	5	9	14	18	23
38	·7813	7841	7869	7898	7926	7954	7983	8012	8040	8069	5	9	14	19	24
39	·8098	8127	8156	8185	8214	8243	8273	8302	8332	8361	5	10	15	20	24
40	·8391	8421	8451	8481	8511	8541	8571	8601	8632	8662	5	10	15	20	25
41	·8693	8724	8754	8785	8816	8847	8878	8910	8941	8972	5	10	16	21	26
42	·9004	9036	9067	9099	9131	9163	9195	9228	9260	9293	5	11	16	21	27
43	·9325	9358	9391	9424	9457	9490	9523	9556	9590	9623	6	11	17	22	28
44	·9657	9691	9725	9759	9793	9827	9861	9896	9930	9965	6	11	17	23	29

NATURAL TANGENTS

Degree	0′	6′	12′	18′	24′	30′	36′	42′	48′	54′	1′	2′	3′	4′	5′
	MINUTES										Mean Differences				
45	1·0000	0035	0070	0105	0141	0176	0212	0247	0283	0319	6	12	18	24	30
46	1·0355	0392	0428	0464	0501	0538	0575	0612	0649	0686	6	12	18	25	31
47	1·0724	0761	0799	0837	0875	0913	0951	0990	1028	1067	6	13	19	25	32
48	1·1106	1145	1184	1224	1263	1303	1343	1383	1423	1463	7	13	20	27	33
49	1·1504	1544	1585	1626	1667	1708	1750	1792	1833	1875	7	14	21	28	34
50	1·1918	1960	2002	2045	2088	2131	2174	2218	2261	2305	7	14	22	29	36
51	1·2349	2393	2437	2482	2527	2572	2617	2662	2708	2753	8	15	23	30	38
52	1·2799	2846	2892	2938	2985	3032	3079	3127	3175	3222	8	16	24	31	39
53	1·3270	3319	3367	3416	3465	3514	3564	3613	3663	3713	8	16	25	33	41
54	1·3764	3814	3865	3916	3968	4019	4071	4124	4176	4229	9	17	26	34	43
55	1·4281	4335	4388	4442	4496	4550	4605	4659	4715	4770	9	18	27	30	45
56	1·4826	4882	4938	4994	5051	5108	5166	5224	5282	5340	10	19	29	38	48
57	1·5399	5458	5517	5577	5637	5697	5757	5818	5880	5941	10	20	30	40	50
58	1·6003	6066	6128	6191	6255	6319	6383	6447	6512	6577	11	21	32	43	53
59	1·6643	6709	6775	6842	6909	6977	7045	7113	7182	7251	11	23	34	45	56
60	1·7321	7391	7461	7532	7603	7675	7747	7820	7893	7966	12	24	36	48	60
61	1·8040	8115	8190	8265	8341	8418	8495	8572	8650	8728	13	26	38	51	64
62	1·8807	8887	8967	9047	9128	9210	9292	9375	9458	9542	14	27	41	55	68
63	1·9626	9711	9797	9883	9970	2·0057	2·0145	2·0233	2·0323	2·0413	15	29	44	58	73
64	2·0503	0594	0686	0778	0872	0965	1060	1155	1251	1348	16	31	47	63	78
65	2·1445	1543	1642	1742	1842	1943	2045	2148	2251	2355	17	34	51	68	85
66	2·2460	2566	2673	2781	2889	2998	3109	3220	3332	3445	18	37	55	73	92
67	2·3559	3673	3789	3906	4023	4142	4262	4383	4504	4627	20	40	60	79	99
68	2·4751	4876	5002	5129	5257	5386	5517	5649	5782	5916	22	43	65	87	108
69	2·6051	6187	6325	6464	6605	6746	6889	7034	7179	7326	24	47	71	95	119
70	2·7475	7625	7776	7929	8083	8239	8397	8556	8716	8878	26	52	78	104	131
71	2·9042	9208	9375	9544	9714	9887	3·0061	3·0237	3·0415	3·0595	29	58	87	116	145
72	3·0777	0961	1146	1334	1524	1716	1910	2106	2305	2506	32	64	96	129	161
73	3·2709	2914	3122	3332	3544	3759	3977	4197	4420	4646	36	72	108	144	180
74	3·4874	5105	5339	5576	5816	6059	6305	6554	6806	7062	41	81	122	163	204
75	3·7321	7583	7848	8118	8391	8667	8947	9232	9520	9812	46	93	139	186	232
76	4·0108	0408	0713	1022	1335	1653	1976	2303	2635	2972	53	107	160	213	267
77	4·3315	3662	4015	4374	4737	5107	5483	5864	6252	6646					
78	4·7046	7453	7867	8288	8716	9152	9594	5·0045	5·0504	5·0970					
79	5·1446	1929	2422	2924	3435	3955	4486	5026	5578	6140					
80	5·6713	7297	7894	8502	9124	9758	6·0405	6·1066	6·1742	6·2432					
81	6·3138	3859	4596	5350	6122	6912	7720	8548	9395	7·0264					
82	7·1154	2066	3002	3962	4947	5958	6996	8062	9158	8·0285					
83	8·1443	2636	3863	5126	6427	7769	9152	9·0579	9·2052	9·3572					
84	9·5144	9·677	9·845	10·02	10·20	10·39	10·58	10·78	10·99	11·20					
85	11·43	11·66	11·91	12·16	12·43	12·71	13·00	13·30	13·62	13·95					
86	14·30	14·67	15·06	15·46	15·89	16·35	16·83	17·34	17·89	18·46					
87	19·08	19·74	20·45	21·20	22·02	22·90	23·86	24·90	26·03	27·27					
88	28·64	30·14	31·82	33·69	35·80	38·19	40·92	44·07	47·74	52·08					
89	57·29	63·66	71·62	81·85	95·49	114·6	143·2	191·0	286·5	573·0					

LOGARITHMIC SINES

Degree	0′	6′	12′	18′	24′	30′	36′	42′	48′	54′	Mean Differences 1′	2′	3′	4′	5′
			M	I	N	U	T	E	S						
0	−∞	$\bar{3}$2419	5429	7190	8439	9408	**0200**	**0870**	**1450**	**1961**					
1	$\bar{2}$·2419	2832	3210	3558	3880	4179	4459	4723	4971	5206					
2	$\bar{2}$·5428	5640	5842	6035	6220	6397	6567	6731	6889	7041					
3	$\bar{2}$·7188	7330	7468	7602	7731	7857	7979	8098	8213	8326					
4	$\bar{2}$·8436	8534	8647	8749	8849	8946	9042	9135	9226	9315	16	32	48	64	80
5	$\bar{2}$·9403	9489	9573	9655	9736	9816	9894	9970	**0046**	**0120**	13	26	39	52	65
6	$\bar{1}$·0192	0264	0334	0403	0472	0539	0605	0670	0734	0797	11	22	33	44	55
7	$\bar{1}$·0859	0920	0981	1040	1099	1157	1214	1271	1326	1381	10	19	29	38	48
8	$\bar{1}$·1436	1489	1542	1594	1646	1697	1747	1797	1847	1895	8	17	25	34	42
9	$\bar{1}$·1943	1991	2038	2085	2131	2176	2221	2266	2310	2353	8	15	23	30	38
10	$\bar{1}$·2397	2439	2482	2524	2565	2606	2647	2687	2727	2767	7	14	20	27	34
11	$\bar{1}$·2806	2845	2883	2921	2959	2997	3034	3070	3107	3143	6	12	19	25	31
12	$\bar{1}$·3179	3214	3250	3284	3319	3353	3387	3421	3455	3488	6	11	27	23	28
13	$\bar{1}$·3521	3554	3586	3618	3650	3682	3713	3745	3775	3806	5	11	16	21	26
14	$\bar{1}$·3837	3867	3897	3927	3957	3986	4015	4044	4073	4102	5	10	15	20	24
15	$\bar{1}$·4130	4158	4186	4214	4242	4269	4296	4323	4350	4377	5	9	14	18	23
16	$\bar{1}$·4403	4430	4456	4482	4508	4533	4559	4584	4609	4634	4	9	13	17	21
17	$\bar{1}$·4659	4684	4709	4733	4757	4781	4805	4829	4853	4876	4	8	12	16	20
18	$\bar{1}$·4900	4923	4946	4969	4992	5015	5037	5060	5082	5104	4	8	11	15	19
19	$\bar{1}$·5126	5148	5170	5192	5213	5235	5256	5278	5299	5320	4	7	11	14	18
20	$\bar{1}$·5341	5361	5382	5402	5423	5443	5463	5484	5504	5523	3	7	10	14	17
21	$\bar{1}$·5543	5563	5583	5602	5621	5641	5660	5679	5698	5717	3	6	10	13	16
22	$\bar{1}$.5736	5754	5773	5792	5810	5828	5847	5865	5883	5901	3	6	9	12	15
23	$\bar{1}$·5919	5937	5954	5972	5990	6007	6024	6042	6059	6076	3	6	9	12	15
24	$\bar{1}$·6093	6110	6127	6144	6161	6177	6194	6210	6227	6243	3	6	8	11	14
25	$\bar{1}$·6259	6276	6292	6308	6324	6340	6356	6371	6387	6403	3	5	8	11	13
26	$\bar{1}$·6418	6434	6449	6465	6480	6495	6510	6526	6541	6556	3	5	8	10	13
27	$\bar{1}$·6570	6585	6600	6615	6629	6644	6659	6673	6687	6702	2	5	7	10	12
28	$\bar{1}$·6716	6730	6744	6759	6773	6783	6801	6814	6828	6842	2	5	7	9	12
29	$\bar{1}$·6856	6869	6883	6896	6910	6923	6937	6950	6963	6977	2	4	7	9	11
30	$\bar{1}$·6990	7003	7016	7029	7042	7055	7068	7080	7093	7106	2	4	6	9	11
31	$\bar{1}$·7118	7131	7144	7156	7168	7181	7193	7205	7218	7230	2	4	6	8	10
32	$\bar{1}$·7242	7254	7266	7278	7290	7302	7314	7326	7338	7349	2	4	6	8	10
33	$\bar{1}$·7361	7373	7384	7396	7407	7419	7430	7442	7453	7464	2	4	6	8	10
34	$\bar{1}$·7476	7487	7498	7509	7520	7531	7542	7553	7564	7575	2	4	6	7	9
35	$\bar{1}$·7586	7597	7607	7618	7629	7640	7650	7661	7671	7682	2	4	5	7	9
36	$\bar{1}$·7692	7703	7713	7723	7734	7744	7754	7764	7774	7785	2	3	5	7	9
37	$\bar{1}$·7795	7805	7815	7825	7835	7844	7854	7864	7874	7884	2	3	5	7	8
38	$\bar{1}$·7893	7903	7913	7922	7932	7941	7951	7960	7970	7979	2	3	5	6	8
39	$\bar{1}$·7989	7998	8007	8017	8026	8035	8044	8053	8063	8072	2	3	5	6	8
40	$\bar{1}$·8081	8090	8099	8108	8117	8125	8134	8143	8152	8161	1	3	4	6	7
41	$\bar{1}$·8169	8178	8187	8195	8204	8213	8221	8230	8238	8247	1	3	4	6	7
42	$\bar{1}$.8255	8264	8272	8280	8289	8297	8305	8313	8322	8330	1	3	4	6	7
43	$\bar{1}$·8338	8346	8354	8362	8370	8378	8386	8394	8402	8410	1	3	4	5	7
44	$\bar{1}$·8418	8426	8433	8441	8449	8457	8464	8472	8480	8487	1	3	4	5	6

LOGARITHMIC SINES

Degree	0′	M 6′	I 12′	N 18′	U 24′	T 30′	E 36′	S 42′	48′	54′	Mean Differences 1′	2′	3′	4′	5′
45	$\bar{1}$·8495	8502	8510	8517	8525	8532	8540	8547	8555	8562	1	2	4	5	6
46	$\bar{1}$·8569	8577	8584	8591	8598	8606	8613	8620	8627	8634	1	2	4	5	6
47	$\bar{1}$·8641	8648	8655	8662	8669	8676	8683	8690	8697	8704	1	2	3	5	6
48	$\bar{1}$·8711	8718	8724	8731	8738	8745	8751	8758	8765	8771	1	2	3	4	6
49	$\bar{1}$·8778	8784	8791	8797	8804	8810	8817	8823	8830	8836	1	2	3	4	5
50	$\bar{1}$·8843	8849	8855	8862	8868	8874	8880	8887	8893	8899	1	2	3	4	5
51	$\bar{1}$·8905	8911	8917	8923	8929	8935	8941	8947	8953	8959	1	2	3	4	5
52	$\bar{1}$·8965	8971	8977	8983	8989	8995	9000	9006	9012	9018	1	2	3	4	5
53	$\bar{1}$·9023	9029	9035	9041	9046	9052	9057	9063	9069	9074	1	2	3	4	5
54	$\bar{1}$·9080	9085	9091	9096	9101	9107	9112	9118	9123	9128	1	2	3	4	5
55	$\bar{1}$·9134	9139	9144	9149	9155	9160	9165	9170	9175	9181	1	2	3	3	4
56	$\bar{1}$·9186	9191	9196	9201	9206	9211	9216	9221	9226	9231	1	2	3	3	4
57	$\bar{1}$·9236	9241	9246	9251	9255	9260	9265	9270	9275	9279	1	2	2	3	4
58	$\bar{1}$·9284	9289	9294	9298	9303	9308	9312	9317	9322	9326	1	2	2	3	4
59	$\bar{1}$·9331	9335	9340	9344	9349	9353	9358	9362	9367	9371	1	1	2	3	4
60	$\bar{1}$·9375	9380	9384	9388	9393	9397	9401	9406	9410	9414	1	1	2	3	4
61	$\bar{1}$·9418	9422	9427	9431	9435	9439	9443	9447	9451	9455	1	1	2	3	3
62	$\bar{1}$·9459	9463	9467	9471	9475	9479	9483	9487	9491	9495	1	1	2	3	3
63	$\bar{1}$·9499	9503	9507	9510	9514	9518	9522	9525	9529	9533	1	1	2	3	3
64	$\bar{1}$·9537	9540	9544	9548	9551	9555	9558	9562	9566	9569	1	1	2	2	3
65	$\bar{1}$·9573	9576	9580	9583	9587	9590	9594	9597	9601	9604	1	1	2	2	3
66	$\bar{1}$·9607	9611	9614	9617	9621	9624	9627	9631	9634	9637	1	1	2	2	3
67	$\bar{1}$·9640	9643	9647	9650	9653	8656	9659	9662	9666	9669	1	1	2	2	3
68	$\bar{1}$·9672	9675	9678	9681	9684	9687	9690	9693	9696	9699	0	1	1	2	2
69	$\bar{1}$·9702	9704	9707	9710	9713	9716	9719	9722	9724	9727	0	1	1	2	2
70	$\bar{1}$·9730	9733	9735	9738	9741	9743	9746	9749	9751	9754	0	1	1	2	2
71	$\bar{1}$·9757	9759	9762	9764	9767	9770	9772	9775	9777	9780	0	1	1	2	2
72	$\bar{1}$·9782	9785	9787	9789	9792	9794	9797	9799	9801	9804	0	1	1	2	2
73	$\bar{1}$·9806	9808	9811	9813	9815	9817	9820	9822	9824	9826	0	1	1	2	2
74	$\bar{1}$·9828	9831	9833	9835	9837	9839	9841	9843	9845	9847	0	1	1	1	2
75	$\bar{1}$·9849	9851	9853	9855	9857	9859	9861	9863	9865	9867	0	1	1	1	2
76	$\bar{1}$·9869	9871	9873	9875	9876	9878	9880	9882	9884	9885	0	1	1	1	2
77	$\bar{1}$·9887	9889	9891	9892	9894	9896	9897	9899	9901	9902	0	1	1	1	1
78	$\bar{1}$·9904	9906	9907	9909	9910	9912	9913	9915	9916	9918	0	1	1	1	1
79	$\bar{1}$·9919	9921	9922	9924	9925	9927	9928	9929	9931	9932	0	0	1	1	1
80	$\bar{1}$·9934	9935	9936	9937	9939	9940	9941	9943	9944	9945	0	0	1	1	1
81	$\bar{1}$·9946	9947	9949	9950	9951	9952	9953	9954	9955	9956	0	0	1	1	1
82	$\bar{1}$·9958	9959	9960	9961	9962	9963	9964	9965	9966	9967	0	0	1	1	1
83	$\bar{1}$·9968	9968	9969	9970	9971	9972	9973	9974	9975	9975	0	0	0	1	1
84	$\bar{1}$·9976	9977	9978	9978	9979	9980	9981	9981	9982	9983	0	0	0	0	1
85	$\bar{1}$·9983	9984	9985	9985	9986	9987	9987	9988	9988	9989	0	0	0	0	0
86	$\bar{1}$·9989	9990	9990	9991	9991	9992	9992	9993	9993	9994	0	0	0	0	0
87	$\bar{1}$·9994	9994	9995	9995	9996	9996	9996	9996	9997	9997	0	0	0	0	0
88	$\bar{1}$·9997	9998	9998	9998	9998	9999	9999	9999	9999	9999	0	0	0	0	0
89	$\bar{1}$·9999	9999	0000	0000	0000	0000	0000	0000	0000	0000	0	0	0	0	0

LOGARITHMIC COSINES

Degree		M	I	N	U	T	E	S		Mean Differences				
	0′	6′	12′	18′	24′	30′	36′	42′	48′	54′	1′	2′	3′	4′	5′
0	$\bar{0}$·0000	0000	0000	0000	0000	0000	0000	0000	0000	$\bar{1}$·9999	0	0	0	0	0
1	$\bar{1}$·9999	9999	9999	9999	9999	9999	9998	9998	9998	9998	0	0	0	0	0
2	$\bar{1}$·9997	9997	9997	9996	9996	9996	9996	9995	9995	9994	0	0	0	0	0
3	$\bar{1}$·9994	9994	9993	9993	9992	9992	9991	9991	9990	9990	0	0	0	0	0
4	$\bar{1}$·9989	9989	9988	9988	9987	9987	9986	9985	9985	9984	0	0	0	0	0
5	$\bar{1}$·9983	9983	9982	9981	9981	9980	9979	9978	9978	9977	0	0	0	0	1
6	$\bar{1}$·9976	9975	9975	9974	9973	9972	9971	9970	9969	9968	0	0	0	1	1
7	$\bar{1}$·9968	9967	9966	9965	9964	9963	9962	9961	9960	9959	0	0	1	1	1
8	$\bar{1}$·9958	9956	9955	9954	9953	9952	9951	9950	9949	9947	0	0	1	1	1
9	$\bar{1}$·9946	9945	9944	9943	9941	9940	9939	9937	9936	9935	0	0	1	1	1
10	$\bar{1}$·9934	9932	9931	9929	9928	9927	9925	9924	9922	9921	0	0	1	1	1
11	$\bar{1}$·9919	9918	9916	9915	9913	9912	9910	9909	9907	9906	0	1	1	1	1
12	$\bar{1}$·9904	9902	9901	9899	9897	9896	9894	9892	9891	9889	0	1	1	1	1
13	$\bar{1}$·9887	9885	9884	9882	9880	9878	9876	9875	9873	9871	0	1	1	1	2
14	$\bar{1}$·9869	9867	9865	9863	9861	9859	9857	9855	9853	9851	0	1	1	1	2
15	$\bar{1}$·9849	9847	9845	9843	9841	9839	9837	9835	9833	9831	0	1	1	1	2
16	$\bar{1}$·9828	9826	9824	9822	9820	9817	9815	9813	9811	9808	0	1	1	2	2
17	$\bar{1}$·9806	9804	9801	9799	9797	9794	9792	9789	9787	9785	0	1	1	2	2
18	$\bar{1}$·9782	9780	9777	9775	9772	9770	9767	9764	9762	9759	0	1	1	2	2
19	$\bar{1}$·9757	9754	9751	9749	9746	9743	9741	9738	9735	9733	0	1	1	2	2
20	$\bar{1}$·9730	9727	9724	9722	9719	9716	9713	9710	9707	9704	0	1	1	2	2
21	$\bar{1}$·9702	9699	9696	9693	9690	9687	9684	9681	9678	9675	0	1	1	2	2
22	$\bar{1}$·9672	9669	9666	9662	9659	9656	9653	9650	9647	9643	1	1	2	2	3
23	$\bar{1}$·9640	9637	9634	9631	9627	9624	9621	9617	9614	9611	1	1	2	2	3
24	$\bar{1}$·9607	9604	9601	9597	9594	9590	9587	9583	9580	9576	1	1	2	2	3
25	$\bar{1}$·9573	9569	9566	9562	9558	9555	9551	9548	9544	9540	1	1	2	2	3
26	$\bar{1}$·9537	9533	9529	9525	9522	9518	9514	9510	9507	9503	1	1	2	3	3
27	$\bar{1}$·9499	9495	9491	9487	9483	9479	9475	9471	9467	9463	1	1	2	3	3
28	$\bar{1}$·9459	9455	9451	9447	9443	9439	9435	9431	9427	9422	1	1	2	3	3
29	$\bar{1}$·9418	9414	9410	9406	9401	9397	9393	9388	9384	9380	1	1	2	3	4
30	$\bar{1}$·9375	9371	9367	9362	9358	9353	9349	9344	9340	9335	1	1	2	3	4
31	$\bar{1}$·9331	9326	9322	9317	9312	9308	9303	9298	9294	9289	1	2	2	3	4
32	$\bar{1}$·9284	9279	9275	9270	9265	9260	9255	9251	9246	9241	1	2	2	3	4
33	$\bar{1}$·9236	9231	9226	9221	9216	9211	9206	9201	9196	9191	1	2	3	3	4
34	$\bar{1}$·9186	9181	9175	9170	9165	9160	9155	9149	9144	9139	1	2	3	3	4
35	$\bar{1}$·9134	9128	9123	9118	9112	9107	9101	9096	9091	9085	1	2	3	4	5
36	$\bar{1}$·9080	9074	9069	9063	9057	9052	9046	9041	9035	9029	1	2	3	4	5
37	$\bar{1}$·9023	9018	9012	9006	9000	8995	8989	8983	8977	8971	1	2	3	4	5
38	$\bar{1}$·8965	8959	8953	8947	8941	8935	8929	8923	8917	8911	1	2	3	4	5
39	$\bar{1}$·8905	8899	8893	8887	8880	8874	8868	8862	8855	8849	1	2	3	4	5
40	$\bar{1}$·8843	8836	8830	8823	8817	8810	8804	8797	8791	8784	1	2	3	4	5
41	$\bar{1}$·8778	8771	8765	8758	8751	8745	8738	8731	8724	8718	1	2	3	5	6
42	$\bar{1}$·8711	8704	8697	8690	8683	8676	8669	8662	8655	8648	1	2	3	5	6
43	$\bar{1}$·8641	8634	8627	8620	8613	8606	8598	8591	8584	8577	1	2	4	5	6
44	$\bar{1}$·8569	8562	8555	8547	8540	8532	8525	8517	8510	8502	1	2	4	5	6

**Subtract mean differences*

LOGARITHMIC COSINES

Degree	MINUTES 0′	6′	12′	18′	24′	30′	36′	42′	48′	54′	Mean Differences 1′	2′	3′	4′	5′
45	$\bar{1}$·8495	8487	8480	8472	8464	8457	8449	8441	8433	8426	1	3	4	5	6
46	$\bar{1}$·8418	8410	8402	8394	8386	8378	8370	8362	8354	8346	1	3	4	5	7
47	$\bar{1}$·8338	8330	8322	8313	8305	8297	8289	8280	8272	8264	1	3	4	6	7
48	$\bar{1}$·8255	8247	8238	8230	8221	8213	8204	8195	8187	8178	1	3	4	6	7
49	$\bar{1}$·8169	8161	8152	8143	8134	8125	8117	8108	8099	8090	1	3	4	6	7
50	$\bar{1}$·8081	8072	8063	8053	8044	8035	8026	8017	8007	7998	2	3	5	6	8
51	$\bar{1}$·7989	7979	7970	7960	7951	7941	7932	7922	7913	7903	2	3	5	6	8
52	$\bar{1}$·7893	7884	7874	7864	7854	7844	7835	7825	7815	7805	2	3	5	7	8
53	$\bar{1}$·7795	7785	7774	7764	7754	7744	7734	7723	7713	7703	2	3	5	7	9
54	$\bar{1}$·7692	7682	7671	7661	7650	7640	7629	7618	7607	7597	2	4	5	7	9
55	$\bar{1}$·7586	7575	7564	7553	7542	7531	7520	7509	7498	7487	2	4	6	7	9
56	$\bar{1}$·7476	7464	7453	7442	7430	7419	7407	7396	7384	7373	2	4	6	8	10
57	$\bar{1}$·7361	7349	7338	7326	7314	7302	7290	7278	7266	7254	2	4	6	8	10
58	$\bar{1}$·7242	7230	7218	7205	7193	7181	7168	7156	7144	7131	2	4	6	8	10
59	$\bar{1}$·7118	7106	7093	7080	7068	7055	7042	7029	7016	7003	2	4	6	9	11
60	$\bar{1}$·6990	6977	6963	6950	6937	6923	6910	6896	6883	6869	2	4	7	9	11
61	$\bar{1}$·6856	6842	6828	6814	6801	6787	6773	6759	6744	6730	2	5	7	9	12
62	$\bar{1}$·6716	6702	6687	6673	6659	6644	6629	6615	6600	6585	2	5	7	10	12
63	$\bar{1}$·6570	6556	6541	6526	6510	6495	6480	6465	6449	6434	3	5	8	10	13
64	$\bar{1}$·6418	6403	6387	6371	6356	6340	6324	6308	6292	6276	3	5	8	11	13
65	$\bar{1}$·6259	6243	6227	6210	6194	6177	6161	6144	6127	6110	3	6	8	11	14
66	$\bar{1}$·6093	6076	6059	6042	6024	6007	5990	5972	5954	5937	3	6	9	12	15
67	$\bar{1}$·5919	5901	5883	5865	5847	5828	5810	5792	5773	5754	3	6	9	12	15
68	$\bar{1}$·5736	5717	5698	5679	5660	5641	5621	5602	5583	5563	3	6	10	13	16
69	$\bar{1}$·5543	5523	5504	5484	5463	5443	5423	5402	5382	5361	3	7	10	14	17
70	$\bar{1}$·5341	5320	5299	5278	5256	5235	5213	5192	5170	5148	4	7	11	14	18
71	$\bar{1}$·5126	5104	5082	5060	5037	5015	4992	4969	4946	4923	4	8	11	15	19
72	$\bar{1}$·4900	4876	4853	4829	4805	4781	4757	4733	4709	4684	4	8	12	16	20
73	$\bar{1}$·4659	4634	4609	4584	4559	4533	4508	4482	4456	4430	4	9	13	17	21
74	$\bar{1}$·4403	4377	4350	4323	4296	4269	4242	4214	4186	4158	5	9	14	18	23
75	$\bar{1}$·4130	4102	4073	4044	4015	3986	3957	3927	3897	3867	5	10	15	20	24
76	$\bar{1}$·3837	3806	3775	3745	3713	3682	3650	3618	3586	3554	5	11	16	21	26
77	$\bar{1}$·3521	3488	3455	3421	3387	3353	3319	3284	3250	3214	6	11	17	23	28
78	$\bar{1}$·3179	3143	3107	3070	3034	2997	2959	2921	2883	2845	6	12	19	25	31
79	$\bar{1}$·2806	2767	2727	2687	2647	2606	2565	2524	2482	2439	7	14	20	27	34
80	$\bar{1}$·2397	2353	2310	2266	2221	2176	2131	2085	2038	1991	8	15	23	30	38
81	$\bar{1}$·1943	1895	1847	1797	1747	1697	1646	1594	1542	1489	8	17	25	34	42
82	$\bar{1}$·1436	1381	1326	1271	1214	1157	1099	1040	0981	0920	10	19	29	38	48
83	$\bar{1}$·0859	0797	0734	0670	0605	0539	0472	0403	0334	0264	11	22	33	44	55
84	$\bar{1}$·0192	0120	0046	**9970**	**9894**	**9816**	**9736**	**9655**	**9573**	**9489**	13	26	39	52	65
85	$\bar{2}$·9403	9315	9226	9135	9042	8946	8849	8749	8647	8543	16	32	48	64	80
86	$\bar{2}$·8436	8326	8213	8098	7979	7857	7731	7602	7468	7330					
87	$\bar{2}$·7188	7041	6889	6731	6567	6397	6220	6035	5842	5640					
88	$\bar{2}$·5428	5206	4971	4723	4459	4179	3880	3558	3210	2832					
89	$\bar{2}$·2419	1961	1450	0870	0200	**9408**	**8439**	**7190**	**5429**	**2419**					

**Subtract mean differences*

LOGARITHMIC TANGENTS

Degree	0′	6′	12′	18′	24′	30′	36′	42′	48′	54′	Mean Differences 1′	2′	3′	4′	5′
			M	I	N	U	T	E	S						
0	— ∞	$\bar{3}$·2419	5429	7190	8439	9409	**0200**	**0870**	**1450**	**1962**					
1	$\bar{2}$·2419	2833	3211	3559	3881	4181	4461	4725	4973	5208					
2	$\bar{2}$·5431	5643	5845	6038	6223	6401	6571	6736	6894	7046					
3	$\bar{2}$·7194	7337	7475	7609	7739	7865	7988	8107	8223	8336					
4	$\bar{2}$·8446	8554	8659	8762	8862	8960	9056	9150	9241	9331	16	32	48	64	81
5	$\bar{2}$·9420	9506	9591	9674	9756	9836	9915	9992	**0069**	**0143**	13	26	40	53	66
6	$\bar{1}$·0216	0289	0360	0430	0499	0567	0633	0699	0764	0828	11	22	34	45	56
7	$\bar{1}$·0891	0954	1015	1076	1135	1194	1252	1310	1367	1423	10	20	29	39	49
8	$\bar{1}$·1478	1533	1587	1640	1693	1745	1797	1848	1898	1948	9	17	26	35	43
9	$\bar{1}$·1997	2046	2094	2142	2189	2236	2282	2328	2374	2419	8	16	23	31	39
10	$\bar{1}$·2463	2507	2551	2594	2637	2680	2722	2764	2805	2846	7	14	21	28	35
11	$\bar{1}$·2887	2927	2967	3006	3046	3085	3123	3162	3200	3237	6	13	19	26	32
12	$\bar{1}$·3275	3312	3349	3385	3422	3458	3493	3529	3564	3599	6	12	18	24	30
13	$\bar{1}$·3634	3668	3702	3736	3770	3804	3837	3870	3903	3935	6	11	17	22	28
14	$\bar{1}$·3968	4000	4032	4064	4095	4127	4158	4189	4220	4250	5	10	16	21	26
15	$\bar{1}$·4281	4311	4341	4371	4400	4430	4459	4488	4517	4546	5	10	15	20	25
16	$\bar{1}$·4575	4603	4632	4660	4688	4716	4744	4771	4799	4826	5	9	14	19	23
17	$\bar{1}$·4853	4880	4907	4934	4961	4987	5014	5040	5066	5092	4	9	13	18	22
18	$\bar{1}$·5118	5143	5169	5195	5220	5245	5270	5295	5320	5345	4	8	13	17	21
19	$\bar{1}$·5370	5294	5419	5443	5467	5491	5516	5539	5563	5587	4	8	12	16	20
20	$\bar{1}$·5611	5634	5658	5681	5704	5727	5750	5773	5796	5819	4	8	12	15	19
21	$\bar{1}$·5842	5864	5887	5909	5932	5954	5976	5998	6020	6042	4	7	11	15	19
22	$\bar{1}$·6064	6086	6108	6129	6151	6172	6194	6215	6236	6257	4	7	11	14	18
23	$\bar{1}$·6279	6300	6321	6341	6362	6383	6404	6424	6445	6465	3	7	10	14	17
24	$\bar{1}$·6486	6506	6527	6547	6567	6587	6607	6627	6647	6667	3	7	10	13	17
25	$\bar{1}$·6687	6706	6726	6746	6765	6785	6804	6824	6843	6863	3	7	10	13	16
26	$\bar{1}$·6882	6901	6920	6939	6958	6977	6996	7015	7034	7053	3	6	9	13	16
27	$\bar{1}$·7072	7090	7109	7128	7146	7165	7183	7202	7220	7238	3	6	9	12	15
28	$\bar{1}$·7257	7275	7293	7311	7330	7348	7366	7384	7402	7420	3	6	9	12	15
29	$\bar{1}$·7438	7455	7473	7491	7509	7526	7544	7562	7579	7597	3	6	9	12	15
30	$\bar{1}$·7614	7632	7649	7667	7684	7701	7719	7736	7753	7771	3	6	9	12	14
31	$\bar{1}$·7788	7805	7822	7839	7856	7873	7890	7907	7924	7941	3	6	9	11	14
32	$\bar{1}$·7958	7975	7992	8008	8025	8042	8059	8075	8092	8109	3	6	8	11	14
33	$\bar{1}$·8125	8142	8158	8175	8191	8208	8224	8241	8257	8274	3	5	8	11	14
34	$\bar{1}$·8290	8306	8323	8339	8355	8371	8388	8404	8420	8426	3	5	8	11	14
35	$\bar{1}$·8452	8468	8484	8501	8517	8533	8549	8565	8581	8597	3	5	8	11	13
36	$\bar{1}$·8613	8629	8644	8660	8676	8692	8708	8724	8740	8755	3	5	8	11	13
37	$\bar{1}$·8771	8787	8803	8818	8834	8850	8865	8881	8897	8912	3	5	8	10	13
38	$\bar{1}$·8928	8944	8959	8975	8990	9006	9022	9037	9053	9068	3	5	8	10	13
39	$\bar{1}$·9084	9099	9115	9130	9146	9161	9176	9192	9207	9223	3	5	8	10	13
40	$\bar{1}$·9238	9254	9269	9284	9300	9315	9330	9346	9361	9376	3	5	8	10	13
41	$\bar{1}$·9392	9407	9422	9438	9453	9468	9483	9499	9514	9529	3	5	8	10	13
42	$\bar{1}$·9544	9560	9575	9590	9605	9621	9636	9651	9666	9681	3	5	8	10	13
43	$\bar{1}$·9697	9712	9727	9742	9757	9773	9788	9803	9818	9833	3	5	8	10	13
44	$\bar{1}$·9848	9864	9879	9894	9909	9924	9939	9955	9970	9985	3	5	8	10	13

LOGARITHMIC TANGENTS

Degree	0′	6′	12′	18′	24′	30′	36′	42′	48′	54′	Mean Differences 1′	2′	3′	4′	5′
45	0·0000	0015	0030	0045	0061	0076	0091	0106	0121	0136	3	5	8	10	13
46	0·0152	0167	0182	0197	0212	0228	0243	0258	0273	0288	3	5	8	10	13
47	0·0303	0319	0334	0349	0364	0379	0395	0410	0425	0440	3	5	8	10	13
48	0·0456	0471	0486	0501	0517	0532	0547	0562	0578	0593	3	5	8	10	13
49	0·0608	0624	0639	0654	0670	0685	0700	0716	0731	0746	3	5	8	10	13
50	0·0762	0777	0793	0808	0824	0839	0854	0870	0885	0901	3	5	8	10	13
51	0·0916	0932	0947	0963	0978	0994	1010	1025	1041	1056	3	5	8	10	13
52	0·1072	1088	1103	1119	1135	1150	1166	1182	1197	1213	3	5	8	10	13
53	0·1229	1245	1260	1276	1292	1308	1324	1340	1356	1371	3	5	8	11	13
54	0·1387	1403	1419	1435	1451	1467	1483	1499	1516	1532	3	5	8	11	13
55	0·1548	1564	1580	1596	1612	1629	1645	1661	1677	1694	3	5	8	11	14
56	0·1710	1726	1743	1759	1776	1792	1809	1825	1842	1858	3	5	8	11	14
57	0·1875	1891	1908	1925	1941	1958	1975	1992	2008	2025	3	6	8	11	14
58	0·2042	2059	2076	2093	2110	2127	2144	2161	2178	2195	3	6	9	11	14
59	0·2212	2229	2247	2264	2281	2299	2316	2333	2351	2368	3	6	9	12	14
60	0·2386	2403	2421	2438	2456	2474	2491	2509	2527	2545	3	6	9	12	15
61	0·2562	2580	2598	2616	2634	2652	2670	2689	2707	2725	3	6	9	12	15
62	0·2743	2762	2780	2798	2817	2835	2854	2872	2891	2910	3	6	9	12	15
63	0·2928	2947	2966	2985	3004	3023	3042	3061	3080	3099	3	6	9	13	16
64	0·3118	3137	3157	3176	3196	3215	3235	3254	3274	3294	3	6	10	13	16
65	0·3313	3333	3353	3373	3393	3413	3433	3453	3473	3494	3	7	10	13	17
66	0·3514	3535	3555	3576	3596	3617	3638	3659	3679	3700	3	7	10	14	17
67	0·3721	3743	3764	3785	3806	3828	3849	3871	3892	3914	4	7	11	14	18
68	0·3936	3958	3980	4002	4024	4046	4068	4091	4113	4136	4	7	11	15	19
69	0·4158	4181	4204	4227	4250	4273	4296	4319	4342	4366	4	8	12	15	19
70	0·4389	4413	4437	4461	4484	4509	4533	4557	4581	4606	4	8	12	16	20
71	0·4630	4655	4680	4705	4730	4755	47 0	4805	4831	4857	4	8	13	17	21
72	0·4882	4908	4934	4960	4986	5013	5039	5066	5093	5120	4	9	13	18	22
73	0·5147	5174	5201	5229	5256	5284	5312	5340	5368	5397	5	9	14	19	23
74	0·5425	5454	5483	5512	5541	5570	5600	5629	5659	5689	5	10	15	20	25
75	0·5719	5750	5780	5811	5842	5873	5905	5936	5968	6000	5	10	16	21	26
76	0·6032	6065	6097	6130	6163	6196	6230	6264	6298	6332	6	11	17	22	28
77	0·6366	6401	6436	6471	6507	6542	6578	6615	6651	6688	6	12	18	24	30
78	0·6725	6763	6800	6838	6877	6915	6954	6994	7033	7073	6	13	19	26	32
79	0·7113	7154	7195	7236	7278	7320	7363	7406	7449	7493	7	14	21	28	35
80	0·7537	7581	7626	7672	7718	7764	7811	7858	7906	7954	8	16	23	31	39
81	0·8003	8052	8102	8152	8203	8255	8307	8360	8413	8467	9	17	26	35	43
82	0·8522	8577	8633	8690	8748	8806	8865	8924	8985	9046	10	20	29	39	49
83	0·9109	9172	9236	9301	9367	9433	9501	9570	9640	9711	11	22	34	45	56
84	0·9784	9857	9932	**0008**	**0085**	**0164**	**0244**	**0326**	**0409**	**0494**	13	26	40	53	66
85	1·0580	0669	0759	0850	0944	1040	1138	1238	1341	1446	16	32	48	64	81
86	1·1554	1664	1777	1893	2012	2135	2261	2391	2525	2663					
87	1·2806	2954	3106	3264	3429	3599	3777	3962	4155	4357					
88	1·4569	4792	5027	5275	5539	5819	6119	6441	6789	7167					
89	1·7581	8038	8550	9130	9800	**0591**	**1561**	**2810**	**4571**	**7581**					

NATURAL LOGARITHMS

	0	1	2	3	4	5	6	7	8	9	1	2	3	4	5	6	7	8	9
											Mean differences								
1·0	0·0000	0100	0198	0296	0392	0488	0583	0677	0770	0862	10	19	29	38	48	57	67	76	86
1·1	0·0953	1044	1133	1222	1310	1398	1484	1570	1655	1740	9	17	26	35	44	52	61	70	78
1·2	0·1823	1906	1989	2070	2151	2231	2311	2390	2469	2546	8	16	24	32	40	48	56	64	72
1·3	0·2624	2700	2776	2852	2927	3001	3075	3148	3221	3293	7	15	22	30	37	44	52	59	67
1·4	0·3365	3436	3507	3577	3646	3716	3784	3853	3920	3988	7	14	21	28	35	41	48	55	62
1·5	0·4055	4121	4187	4253	4318	4383	4447	4511	4574	4637	6	13	19	26	32	39	45	52	58
1·6	0·4700	4762	4824	4886	4947	5008	5068	5128	5188	5247	6	12	18	24	30	36	42	48	55
1·7	0·5306	5365	5423	5481	5539	5596	5653	5710	5766	5822	6	11	17	23	29	34	40	46	51
1·8	0·5878	5933	5988	6043	6098	6152	6206	6259	6313	6366	5	11	16	22	27	32	38	43	49
1·9	0·6419	6471	6523	6575	6627	6678	6729	6780	6831	6881	5	10	15	20	26	30	36	41	46
2·0	0·6931	6981	7031	7080	7129	7178	7227	7275	7324	7372	5	10	15	20	24	29	34	39	44
2·1	0·7419	7467	7514	7561	7608	7655	7701	7747	7793	7839	5	9	14	19	23	28	33	37	42
2·2	0·7885	7930	7975	8020	8065	8109	8154	8198	8242	8286	4	9	13	18	22	27	31	36	40
2·3	0·8329	8372	8416	8459	8502	8544	8587	8628	8671	8713	4	9	13	17	21	26	30	34	38
2·4	0·8755	8796	8838	8879	8920	8961	9002	9042	9083	9122	4	8	12	16	20	24	29	33	37
2·5	0·9163	9203	9243	9282	9322	9361	9400	9439	9478	9517	4	8	12	16	20	24	27	31	35
2·6	0·9555	9594	9632	9670	9708	9746	9783	9821	9858	9895	4	8	11	15	19	23	26	30	34
2·7	0·9933	9969	**0006**	**0043**	**0080**	**0116**	**0152**	**0188**	**0225**	**0260**	4	7	11	15	18	22	25	29	33
2·8	1·0296	0332	0367	0403	0438	0473	0508	0543	0578	0613	4	7	11	14	18	21	25	28	32
2·9	1·0647	0682	0716	0750	0784	0818	0852	0886	0919	0953	3	7	10	14	17	20	24	27	31
3·0	1·0986	1019	1053	1086	1119	1151	1184	1217	1249	1282	3	7	10	13	16	20	23	26	30
3·1	1·1314	1346	1378	1410	1442	1474	1506	1537	1568	1600	3	6	10	13	16	19	22	25	29
3·2	1·1632	1663	1694	1725	1756	1787	1817	1848	1878	1909	3	6	9	12	15	18	22	25	28
3·3	1·1939	1969	2000	2030	2060	2080	2119	2149	2178	2208	3	6	9	12	15	18	21	24	27
3·4	1·2238	2267	2296	2326	2355	2384	2413	2442	2470	2499	3	6	9	12	15	17	20	23	26
3·5	1·2528	2556	2585	2613	2641	2669	2698	2726	2754	2782	3	6	8	11	14	17	20	23	25
3·6	1·2809	2837	2865	2892	2920	2947	2975	3002	3029	3056	3	5	8	11	14	16	19	22	25
3·7	1·3083	3110	3137	3164	3191	3218	3244	3271	3297	3324	3	5	8	11	13	16	19	21	24
3·8	1·3350	3376	3403	3429	3455	3481	3507	3533	3558	3584	3	5	8	10	13	16	18	21	23
3·9	1·3610	3635	3661	3686	3712	3737	3762	3788	3813	3838	3	5	8	10	13	15	18	20	23
4·0	1·3863	3888	3913	3938	3962	3987	4012	4036	4061	4085	2	5	7	10	12	15	17	20	22
4·1	1·4110	4134	4159	4183	4207	4231	4255	4279	4303	4327	2	5	7	10	12	14	17	19	22
4·2	1·4351	4374	4398	4422	4446	4469	4493	4516	4540	4563	2	5	7	9	12	14	16	19	21
4·3	1·4586	4609	4633	4656	4679	4702	4725	4748	4770	4793	2	5	7	9	12	14	16	18	21
4·4	1·4816	4839	4861	4884	4907	4929	4953	4974	4996	5019	2	5	7	9	11	14	16	18	20
4·5	1·5041	5063	5085	5107	5129	5151	5173	5195	5217	5239	2	4	7	9	11	13	15	18	20
4·6	1·5261	5282	5304	5326	5347	5369	5390	5412	5433	5454	2	4	6	9	11	13	15	17	19
4·7	1·5476	5496	5518	5539	5560	5581	5602	5623	5644	5665	2	4	6	8	11	13	15	17	19
4·8	1·5686	5707	5728	5748	5769	5790	5810	5831	5851	5872	2	4	6	8	10	12	14	16	19
4·9	1·5892	5913	5933	5953	5974	5994	6014	6034	6054	6074	2	4	6	8	10	12	14	16	18
5·0	1·6094	6114	6134	6154	6174	6194	6214	6233	6253	6273	2	4	6	8	10	12	14	16	18
5·1	1·6292	6312	6332	6351	6371	6390	6409	6429	6448	6467	2	4	6	8	10	12	14	16	18
5·2	1·6487	6505	6525	6544	6563	6582	6601	6620	6639	6658	2	4	6	8	10	11	13	15	17
5·3	1·6677	6696	6715	6734	6752	6771	6789	6808	6827	6845	2	4	6	7	9	11	13	15	17
5·4	1·6864	6882	6901	6919	6938	6956	6974	6993	7011	7029	2	4	5	7	9	11	13	15	17

NATURAL LOGARITHMS

	0	1	2	3	4	5	6	7	8	9	Mean differences 1	2	3	4	5	6	7	8	9
5·5	1·7047	7066	7084	7102	7120	7138	7156	7174	7192	7210	2	4	5	7	9	11	13	14	16
5·6	1·7228	7246	7263	7281	7299	7317	7334	7352	7370	7387	2	4	5	7	9	11	12	14	16
5·7	1·7405	7422	7440	7457	7475	7492	7509	7527	7544	7561	2	3	5	7	9	10	12	14	16
5·8	1·7579	7596	7613	7630	7647	7664	7681	7699	7716	7733	2	3	5	7	9	10	12	14	15
5·9	1·7750	7766	7783	7800	7817	7834	7851	7867	7884	7901	2	3	5	7	8	10	12	13	15
6·0	1·7918	7934	7951	7967	7984	8001	8017	8034	8050	8066	2	3	5	7	8	10	12	13	15
6·1	1·8083	8099	8116	8133	8148	8165	8181	8197	8213	8229	2	3	5	6	8	10	11	13	15
6·2	1·8245	8262	8278	8294	8310	8326	8342	8358	8374	8390	2	3	5	6	8	10	11	13	14
6·3	1·8405	8421	8437	8453	8469	8485	8500	8516	8532	8547	2	3	5	6	8	9	11	13	14
6·4	1·8563	8578	8594	8610	8625	8641	8656	8672	8687	8703	2	3	5	6	8	9	11	12	14
6·5	1·8718	8733	8749	8764	8779	8795	8810	8825	8840	8856	2	3	5	6	8	9	11	12	14
6·6	1·8871	8886	8901	8916	8931	8946	8961	8976	8991	9006	2	3	5	6	8	9	11	12	14
6·7	1·9021	9036	9051	9066	9081	9095	9110	9125	9140	9155	1	3	4	6	7	9	10	12	13
6·8	1·9169	9184	9199	9213	9228	9242	9257	9272	9286	9301	1	3	4	6	7	9	10	12	13
6·9	1·9315	9330	9344	9359	9373	9387	9402	9416	9430	9445	1	3	4	6	7	9	10	12	13
7·0	1·9459	9473	9488	9502	9516	9530	9544	9559	9573	9587	1	3	4	6	7	9	10	11	13
7·1	1·9601	9615	9629	9643	9657	9671	9685	9699	9713	9727	1	3	4	6	7	8	10	11	13
7·2	1·9741	9755	9769	9782	9796	9810	9824	9838	9851	9865	1	3	4	6	7	8	10	11	12
7·3	1·9879	9892	9906	9920	9933	9947	9961	9974	9988	**0001**	1	3	4	5	7	8	10	11	12
7·4	2·0015	0028	0042	0055	0069	0082	0096	0109	0122	0136	1	3	4	5	7	8	9	11	12
7·5	2·0149	0162	0176	0189	0202	0215	0229	0242	0255	0268	1	3	4	5	7	8	9	11	12
7·6	2·0281	0295	0308	0321	0334	0347	0360	0373	0386	0399	1	3	4	5	7	8	9	10	12
7·7	2·0412	0425	0438	0451	0464	0477	0490	0503	0516	0528	1	3	4	5	6	8	9	10	12
7·8	2·0541	0554	0567	0580	0592	0605	0618	0631	0643	0656	1	3	4	5	6	8	9	10	11
7·9	2·0668	0681	0694	0707	0719	0732	0744	0757	0769	0782	1	3	4	5	6	8	9	10	11
8·0	2·0794	0807	0719	0832	0844	0857	0869	0882	0894	0906	1	3	4	5	6	7	9	10	11
8·1	2·0919	0931	0943	0956	0968	0980	0992	1005	1017	1029	1	2	4	5	6	7	9	10	11
8·2	2·1041	1054	1066	1078	1090	1102	1114	1126	1138	1150	1	2	4	5	6	7	9	10	11
8·3	2·1163	1175	1187	1199	1211	1223	1235	1247	1258	1270	1	2	4	5	6	7	8	10	11
8·4	2·1282	1294	1306	1318	1330	1342	1353	1365	1377	1389	1	2	4	5	6	7	8	10	11
8·5	2·1401	1412	1424	1436	1448	1459	1471	1483	1494	1506	1	2	4	5	6	7	8	9	11
8·6	2·1518	1529	1541	1552	1564	1576	1587	1599	1610	1622	1	2	3	5	6	7	8	9	10
8·7	2·1633	1645	1656	1668	1679	1691	1702	1713	1725	1736	1	2	3	5	6	7	8	9	10
8·8	2·1748	1759	1770	1782	1793	1804	1815	1827	1838	1849	1	2	3	5	6	7	8	9	10
8·9	2·1861	1871	1883	1894	1905	1917	1928	1939	1950	1961	1	2	3	4	6	7	8	9	10
9·0	2·1972	1983	1994	2006	2017	2028	2039	2050	2061	2072	1	2	3	4	6	7	8	9	10
9·1	2·2083	2094	2105	2116	2127	2138	2148	2159	2170	2181	1	2	3	4	5	7	8	9	10
9·2	2·2192	2203	2214	2225	2235	2246	2257	2268	2279	2289	1	2	3	4	5	6	8	9	10
9·3	2·2300	2311	2322	2332	2343	2354	2364	2375	2386	2396	1	2	3	4	5	6	8	9	10
9·4	2·2407	2418	2428	2439	2450	2460	2471	2481	2492	2502	1	2	3	4	5	6	7	9	10
9·5	2·2513	2523	2534	2544	2555	2565	2576	2586	2597	2607	1	2	3	4	5	6	7	8	10
9·6	2·2618	2628	2638	2649	2659	2670	2680	2690	2701	2711	1	2	3	4	5	6	7	8	9
9·7	2·2721	2732	2742	2752	2762	2773	2783	2793	2803	2814	1	2	3	4	5	6	7	8	9
9·8	2·2824	2834	2844	2854	2865	2875	2885	2895	2905	2915	1	2	3	4	5	6	7	8	9
9·9	2·2925	2935	2946	2956	2966	2976	2986	2996	3006	3016	1	2	3	4	5	6	7	8	9

EXPONENTIAL AND HYPERBOLIC FUNCTIONS

x	e^x	e^{-x}	$\theta°$ (gd x)	cosh x (secθ)	sinh x (tan θ)	tanh x (sin θ)	log cosh x	log sinh x
·1	1·1052	·9048	5·720	1·0050	·1002	·0997	·0022	$\bar{1}$·0007
·2	1·2214	·8187	11·384	1·0201	·2013	·1974	·0086	$\bar{1}$·3039
·3	1·3499	·7408	16·937	1·0453	·3045	·2913	·0193	$\bar{1}$·4836
·4	1·4918	·6703	22·331	1·0811	·4108	·3799	·0339	$\bar{1}$·6136
·5	1·6487	·6065	27·524	1·1276	·5211	·4621	·0522	$\bar{1}$·7169
·6	1·8221	·5488	32·483	1·1855	·6367	·5370	·0739	$\bar{1}$·8093
·7	2·0138	·4966	37·183	1·2552	·7586	·6044	·0987	$\bar{1}$·8800
·8	2·2255	·4493	41·608	1·3374	·8881	·6640	·1263	$\bar{1}$·9485
·9	2·4596	·4066	45·750	1·4331	1·0265	·7163	·1563	0·0114
1·0	2·7183	·3679	49·605	1·5431	1·1752	·7616	·1884	0·0701
1·1	3·0042	·3329	53·178	1·6685	1·3356	·8005	·2223	0·1257
1·2	3·3201	·3012	56·476	1·8107	1·5095	·8337	·2578	0·1788
1·3	3·6693	·2725	59·511	1·9709	1·6984	·8617	·2947	0·2300
1·4	4·0552	·2466	62·295	2·1509	1·9043	·8854	·3326	0·2797
1·5	4·4817	·2231	64·843	2·3524	2·1293	·9051	·3715	0·3282
1·6	4·9530	·2019	67·171	2·5775	2·3756	·9217	·4112	0·3758
1·7	5·4739	·1827	69·294	2·8283	2·6456	·9354	·4515	0·4225
1·8	6·0496	·1653	71·228	3·1075	2·9422	·9468	·4924	0·4687
1·9	6·6859	·1496	72·987	3·4177	3·2682	·9562	·5337	0·5143
2·0	7·3891	·1353	74·584	3·7622	3·6269	·9640	·5754	0·5595
2·1	8·1662	·1225	76·037	4·1443	4·0219	·9705	·6175	0·6044
2·2	9·0250	·1108	77·354	4·5679	4·4571	·9757	·6597	0·6491
2·3	9·9742	·1003	78·549	5·0372	4·9370	·9801	·7022	0·6935
2·4	11·023	·0907	79·633	5·5569	5·4662	·9837	·7448	0·7377
2·5	12·183	·0821	80·615	6·1323	6·0502	·9866	·7876	0·7818
2·6	13·464	·0743	81·513	6·7690	6·6947	·9890	·8305	0·8257
2·7	14·880	·0672	82·310	7·4735	7·4063	·9910	·8735	0·8696
2·8	16·445	·0608	83·040	8·2527	8·1919	·9926	·9166	0·9134
2·9	18·174	·0550	83·707	9·1146	9·0596	·9940	·9597	0·9571
3·0	20·086	·0498	84·301	10·068	10·018	·9951	1·0029	1·0008
3·1	22·198	·0450	84·841	11·121	11·076	·9959	1·0462	1·0444
3·2	24·533	·0408	85·336	12·287	12·246	·9967	1·0894	1·0880
3·3	27·113	·0369	85·775	13·575	13·538	·9973	1·1327	1·1316
3·4	29·964	·0334	86·177	14·999	14·965	·9978	1·1761	1·1751
3·5	33·115	·0302	86·541	16·573	16·543	·9982	1·2194	1·2186
3·6	36·598	·0273	86·870	18·313	18·285	·9985	1·2628	1·2621
3·7	40·447	·0247	87·168	20·236	20·211	·9988	1·3061	1·3056
3·8	44·701	·0224	87·437	22·362	22·339	·9990	1·3495	1·3491
3·9	49·402	·0202	87·681	24·711	24·691	·9992	1·3929	1·3925
4·0	54·598	·0183	87·903	27·308	27·290	·9993	1·4363	1·4360
4·1	60·340	·0166	88·104	30·178	30·162	·9995	1·4797	1·4795
4·2	66·686	·0150	88·281	33·351	33·336	·9996	1·5231	1·5229
4·3	73·700	·0136	88·447	36·857	36·843	·9996	1·5665	1·5664
4·4	81·451	·0123	88·591	40·732	40·719	·9997	1·6099	1·6098
4·5	90·017	·0111	88·728	45·014	45·003	·9997	1·6533	1·6532
4·6	99·484	·0101	88·849	49·747	49·737	·9998	1·6968	1·6967
4·7	109·95	·0091	88·957	54·978	54·969	·9998	1·7402	1·7401
4·8	121·51	·0082	89·055	60·759	60·751	·9999	1·7836	1·7836
4·9	134·29	·0074	89·146	67·149	67·141	·9999	1·8270	1·8270
5·0	148·41	·0067	89·227	74·210	74·203	·9999	1·8705	1·8704

VALUES OF GAMMA FUNCTIONS

x	0	1	2	3	4	5	6	7	8	9
1·0	1·0000	0·9943	0·9888	0·9835	0·9784	0·9735	0·9687	0·9641	0·9597	0·9554
1·1	0·9513	0·9474	0·9436	0·9399	0·9364	0·9330	0·9298	0·9267	0·9237	0·9209
1·2	0·9182	0·9156	0·9131	0·9107	0·9085	0·9064	0·9044	0·9025	0·9007	0·8990
1·3	0·8975	0·8960	0·8946	0·8934	0·8922	0·8911	0·8902	0·8893	0·8885	0·8878
1·4	0·8873	0·8868	0·8864	0·8860	0·8858	0·8857	0·8856	0·8856	0·8857	0·8859
1·5	0·8862	0·8866	0·8870	0·8876	0·8882	0·8889	0·8896	0·8905	0·8914	0·8924
1·6	0·8935	0·8947	0·8959	0·8972	0·8986	0·9001	0·9017	0·9033	0·9050	0·9068
1·7	0·9086	0·9106	0·9126	0·9147	0·9168	0·9191	0·9214	0·9238	0·9262	0·9288
1·8	0·9314	0·9341	0·9368	0·9397	0·9426	0·9456	0·9487	0·9518	0·9551	0·9584
1·9	0·9618	0·9652	0·9688	0·9724	0·9761	0·9799	0·9837	0·9877	0·9917	0·9958
2·0	1·0000									

LEGENDRE'S POLYNOMIALS (Spherical functions)

$P_0(x) = 1$; $P_1(x) = x$; $P_2(x) = \frac{1}{2}(3x^2 - 1)$; $P_3(x) = \frac{1}{2}(5x^3 - 3x)$;
$P_4(x) = \frac{1}{8}(35x^4 - 30x^2 + 3)$; $P_5(x) = \frac{1}{8}(63x^5 - 70x^3 + 15x)$;
$P_6(x) = \frac{1}{16}(231x^6 - 315x^4 + 105x^2 - 5)$; $P_7(x) = \frac{1}{16}(429x^7 - 693x^5 + 315x^3 - 35x)$.

$$P_n(x) = \frac{1}{2^n n!} \frac{d^n}{dx^n}[(x^2-1)^n]$$ ....*General Form*

x	$P_0(x)$	$P_1(x)$	$P_2(x)$	$P_3(x)$	$P_4(x)$	$P_5(x)$	$P_6(x)$	$P_7(x)$
0·00	1	0·00	—0·5000	0·0000	0·3750	0·0000	0·3125	0·0000
0·05	1	0·05	—0·4962	—0·0747	0·3657	0·0927	—0·2962	—0·1069
0·10	1	0·10	—0·4850	—0·1475	0·3379	0·1788	—0·2488	—0·1995
0·15	1	0·15	—0·4662	—0·2166	0·2928	0·2523	—0·1746	—0·2649
0·20	1	0·20	—0·4400	—0·2800	0·2320	0·3075	—0·0806	—0·2935
0·25	1	0·25	—0·4062	—0·3359	0·1577	0·3397	+0·0243	—0·2799
0·30	1	0·30	—0·3650	—0·3825	+0·0729	0·3454	0·1292	—0·2241
0·35	1	0·35	—0·3162	—0·4178	—0·0187	0·3225	0·2225	—0·1318
0·40	1	0·40	—0·2600	—0·4400	—0·1130	0·2706	0·2926	—0·0146
0·45	1	0·45	—0·1962	—0·4472	—0·2050	0·1917	0·3290	0·1106
0·50	1	0·50	—0·1250	—0·4375	—0·2891	+0·0898	0·3232	0·2231
0·55	1	0·55	—0·0462	—0·4091	—0·3590	—0·0282	0·2708	0·3007
0·60	1	0·60	+0·0400	—0·3600	—0·4080	—0·1526	0·1721	0·3226
0·65	1	0·65	0·1338	—0·2884	—0·4284	—0·2705	0·0347	0·2737
0·70	1	0·70	0·2350	—0·1925	—0·4121	—0·3652	—0·1253	0·1502
0·75	1	0·75	0·3438	—0·0703	—0·3501	—0·4164	—0·2808	—0·0342
0·80	1	0·80	0·4600	+0·0800	—0·2330	—0·3995	—0·3918	—0·2397
0·85	1	0·85	0·5838	0·2603	—0·0506	—0·2857	—0·4030	—0·3913
0·90	1	0·90	0·7150	0·4725	+0·2079	—0·0411	—0·2412	—0·3678
0·95	1	0·95	0·8538	0·7184	0·5541	0·3727	0·1875	0·0112
1·00	1	1·00	1·0000	1·0000	1·0000	1·0000	1·0000	1·0000

HYPERBOLIC FUNCTIONS

x	e^x	e^{-x}	sinh$x$	cosh$x$	tanh$x$	x	e^x	e^{-x}	sinh$x$	cosh$x$	tanh$x$
0·00	1·0000	1·0000	0·0000	1·0000	0·0000						
01	1·0101	0·9901	0·0100	1·0001	0·0100	0·51	1·6653	0·6005	0·5324	1·1329	0·4700
02	1·0202	0·9802	0·0200	1·0002	0·0200	52	1·6820	0·5945	0·5438	1·1383	0·4777
03	1·0305	0·9705	0·0300	1·0005	0·0300	53	1·6989	0·5886	0·5552	1·1438	0·4854
04	1·0408	0·9608	0·0400	1·0008	0·0400	54	1·7160	0·5828	0·5666	1·1494	0·4930
05	1·0513	0·9512	0·0500	1·0013	0·0500	55	1·7333	0·5770	0·5781	1·1551	0·5005
06	1·0618	0·9418	0·0600	1·0018	0·0599	56	1·7507	0·5712	0·5897	1·1609	0·5080
07	1·0725	0·9324	0·0701	1·0025	0·0699	57	1·7683	0·5655	0·6014	1·1669	0·5154
08	1·0833	0·9231	0·0801	1·0032	0·0798	58	1·7860	0·5599	0·6131	1·1730	0·5227
09	1·0942	0·9139	0·0901	1·0041	0·0898	59	1·8040	0·5543	0·6248	1·1792	0·5299
0·10	1·1052	0·9048	0·1002	1·0050	0·0997	0·60	1·8221	0·5488	0·6367	1·1855	0·5371
11	1·1163	0·8958	0·1102	1·0061	0·1096	61	1·8404	0·5434	0·6485	1·1919	0·5441
12	1·1275	0·8869	0·1203	1·0072	0·1194	62	1·8589	0·5379	0·6605	1·1984	0·5511
13	1·1388	0·8781	0·1304	1·0085	0·1293	63	1·8776	0·5326	0·6725	1·2051	0·5581
14	1·1503	0·8694	0·1405	1·0098	0·1391	64	1·8965	0·5273	0·6846	1·2119	0·5649
15	1·1618	0·8607	0·1506	1·0113	0·1489	65	1·9155	0·5221	0·6968	1·2188	0·5717
16	1·1735	0·8521	0·1607	1·0128	0·1587	66	1·9348	0·5169	0·7090	1·2252	0·5784
17	1·1853	0·8437	0·1708	1·0145	0·1684	67	1·9542	0·5117	0·7213	1·2330	0·5850
18	1·1972	0·8353	0·1810	1·0162	0·1781	68	1·9739	0·5066	0·7336	1·2403	0·5915
19	1·2093	0·8270	0·1912	1·0181	0·1878	69	1·9937	0·5016	0·7461	1·2477	0·5980
0·20	1·2214	0·8187	0·2013	1·0201	0·1974	0·70	2·0138	0·4966	0·7586	1·2552	0·6044
21	1·2337	0·8106	0·2116	1·0221	0·2068	71	2·0340	0·4916	0·7712	1.2628	0·6107
22	1·2461	0·8025	0·2218	1·0243	0·2165	72	2·0544	0·4868	0·7838	1·2706	0·6169
23	1·2586	0·7945	0·2320	1·0266	0·2260	73	2·0751	0·4819	0·7966	1·2785	0·6231
24	1·2713	0·7866	0·2423	1·0289	0·2355	74	2·0959	0·4771	0·8094	1·2865	0·6292
25	1·2840	0·7788	0·2526	1·0314	0·2449	75	2·1170	0·4724	0·8223	1·2947	0·6352
26	1·2969	0·7711	0·2629	1·0340	0·2543	76	2·1383	0·4677	0·8353	1·3030	0·6411
27	1·3100	0·7634	0·2733	1·0367	0·2636	77	2·1598	0·4630	0·8484	1·3114	0·6469
28	1·3231	0·7558	0·2837	1·0395	0·2729	78	2·1815	0·4584	0·8615	1·3199	0·6527
29	1·3364	0·7483	0·2941	1·0424	0·2821	79	2·2034	0·4538	0·8748	1·3286	0·6584
0·30	1·3499	0·7408	0·3045	1·0453	0·2913	0·80	2·2255	0·4493	0·8881	1·3374	0·6640
31	1·3634	0·7335	0·3150	1·0484	0·3004	81	2·2479	0·4449	0·9015	1·3464	0·6696
32	1·3771	0·7262	0·3255	1·0516	0·3095	82	2·2705	0·4404	0·9150	1·3555	0·6751
33	1·3910	0·7189	0·3360	1·0550	0·3185	83	2·2933	0·4361	0·9286	1·3647	0·6805
34	1·4050	0·7118	0·3466	1·0584	0·3275	84	2·3164	0·4317	0·9423	1·3740	0·6858
35	1·4191	0·7047	0·3572	1·0619	0·3364	85	2·3397	0·4274	0·9561	1·3835	0·6911
36	1·4333	0·6977	0·3678	1·0655	0·3452	86	2·3632	0·4232	0·9700	1·3932	0·6963
37	1·4477	0·6907	0·3785	1·0692	0·3540	87	2·3869	0·4190	0·9840	1·4029	0·7014
38	1·4623	0·6839	0·3892	1·0731	0·3627	88	2·4109	0·4148	0·9981	1·4128	0·7064
39	1·4770	0·6771	0·4000	1·0770	0·3714	89	2·4351	0·4107	1·0122	1·4229	0·7114
0·40	1·4918	0·6703	0·4108	1·0811	0·3800	0·90	2·4596	0·4066	1·0265	1·4331	0·7163
41	1·5068	0·6637	0·4216	1·0852	0·3885	91	2·4843	0·4025	1·0409	1·4434	0·7211
42	1·5220	0·6571	0·4325	1·0895	0·3969	92	2·5093	0·3985	1·0554	1·4539	0·7259
43	1·5373	0·6505	0·4434	1·0939	0·4053	93	2·5345	0·3946	1·0700	1·4645	0·7306
44	1·5527	0·6440	0·4543	1·0984	0·4136	94	2·5600	0·3906	1·0847	1·4753	0·7352
45	1·5683	0·6376	0·4653	1·1030	0·4219	95	2·5857	0·3867	1·0995	1·4862	0·7398
46	1·5841	0·6313	0·4764	1·1077	0·4301	96	2·6117	0·3829	1·1144	1·4973	0·7443
47	1·6000	0·6250	0·4875	1·1125	0·4382	97	2·6379	0·3791	1·1294	1·5085	0·7487
48	1·6161	0·6188	0·4987	1·1174	0·4462	98	2·6645	0·3753	1·1446	1·5199	0·7531
49	1·6323	0·6126	0·5098	1·1225	0·4542	99	2·6912	0·3716	1·1598	1·5314	0·7574
0·50	1·6487	0·6065	0·5211	1·1276	0·4621	1·00	2·7183	0·3679	1·1752	1·5431	0·7616

HYPERBOLIC FUNCTIONS

x	e^x	e^{-x}	sinh$x$	cosh$x$	tanh$x$	x	e^x	e^{-x}	sinh$x$	cosh$x$	tanh$x$
1·01	2·7456	0·3642	1·1907	1·5549	0·7658	1·51	4·5267	0·2209	2·1529	2·3738	0·9069
02	2·7732	0·3606	1·2063	1·5669	0·7699	52	4·5722	0·2187	2·1768	2·3955	0·9087
03	2·8011	0·3570	1·2220	1·5790	0·7739	53	4·6182	0·2165	2·2008	2·4174	0·9104
04	2·8292	0·3535	1·2379	1·5913	0·7779	54	4·6646	0·2144	2·2251	2·4395	0·9121
05	2·8577	0·3499	1·2539	1·6038	0·7818	55	4·7115	0·2123	2·2496	2·4619	0·9138
06	2·8864	0·3465	1·2700	1·6164	0·7857	56	4·7588	0·2101	2·2743	2·4845	0·9154
07	2·9154	0·3430	1·2862	1·6292	0·7895	57	4·8067	0·2081	2·2993	2·5074	0·9170
08	2·9447	0·3396	1·3025	1·6421	0·7932	58	4·8550	0·2060	2·3245	2·5305	0·9186
09	2·9743	0·3362	1·3190	1·6553	0·7969	59	4·9038	0·2039	2·3499	2·5538	0·9202
1·10	3·0042	0·3329	1·3357	1·6685	0·8005	1·60	4·9530	0·2019	2·3756	2·5775	0·9217
11	3·0344	0·3296	1·3524	1·6820	0·8041	61	5·0028	0·1999	2·4015	2·6014	0·9232
12	3·0649	0·3263	1·3693	1·6956	0·8076	62	5·0531	0·1979	2·4276	2·6255	0·9246
13	3·0957	0·3230	1·3863	1·7093	0·8110	63	5·1039	0·1959	2·4540	2·6499	0·9261
14	3·1268	0·3198	1·4035	1·7233	0·8144	64	5·1552	0·1940	2·4806	2·6746	0·9275
15	3·1582	0·3166	1·4208	1·7374	0·8178	65	5·2070	0·1920	2·5075	2·6995	0·9289
16	3·1899	0·3135	1·4382	1·7517	0·8210	66	5·2593	0·1901	2·5346	2·7247	0·9302
17	3·2220	0·3104	1·4558	1·7662	0·8243	67	5·3122	0·1883	2·5620	2·7502	0·9316
18	3·2544	0·3073	1·4736	1·7808	0·8275	68	5·3656	0·1864	2·5896	2·7760	0·9329
19	3·2871	0·3042	1·4914	1·7957	0·8306	69	5·4195	0·1845	2·6175	2·8020	0·9342
1·20	3·3201	0·3012	1·5095	1·8107	0·8337	1·70	5·4740	0·1827	2·6456	2·8283	0·9354
21	3·3535	0·2982	1·5276	1·8258	0·8367	71	5·5290	0·1809	2·6741	2·8549	0·9367
22	3·3872	0·2952	1·5460	1·8412	0·8397	72	5·5845	0·1791	2·7027	2·8818	0·9379
23	3·4212	0·2923	1·5645	1·8568	0·8426	73	5·6407	0·1773	2·7317	2·9090	0·9391
24	3·4556	0·2894	1·5831	1·8725	0·8455	74	5·6973	0·1755	2·7609	2·9364	0·9402
25	3·4903	0·2865	1·6019	1·8884	0·8483	75	5·7546	0·1738	2·7904	2·9642	0·9414
26	3·5254	0·2837	1·6209	1·9045	0·8511	76	5·8124	0·1720	2·8202	2·9922	0·9425
27	3·5609	0·2808	1·6400	1·9208	0·8538	77	5·8709	0·1703	2·8503	3·0206	0·9436
28	3·5966	0·2780	1·6593	1·9373	0·8565	78	5·9299	0·1686	2·8806	3·0493	0·9447
29	3·6328	0·2753	1·6788	1·9540	0·8591	79	5·9895	0.1670	2·9113	3·0782	0·9458
1·30	3·6693	0·2725	1·6984	1·9709	0·8617	1·80	6·0497	0·1653	2·9422	3·1075	0·9468
31	3·7062	0·2698	1·7182	1·9880	0·8643	81	6·1105	0·1637	2·9734	3·1371	0·9478
32	3·7434	0·2671	1·7381	2·0053	0·8668	82	6·1719	0·1620	3·0049	3·1669	0·9488
33	3·7810	0·2645	1·7583	2·0228	0·8693	83	6·2339	0·1604	3·0367	3·1972	0·9498
34	3·8190	0·2619	1·7786	2·0404	0·8717	84	6·2965	0·1588	3·0689	3·2277	0·9508
35	3·8574	0·2592	1·7991	2·0583	0·8741	85	6·3598	0·1572	3·1013	3·2585	0·9518
36	3·8962	0·2567	1·8198	2·0764	0·8764	86	6·4237	0·1557	3·1340	3·2897	0·9527
37	3·9354	0·2541	1·8406	2·0947	0·8787	87	6·4883	0·1541	3·1671	3·3212	0·9536
38	3·9749	0·2516	1·8617	2·1132	0·8810	88	6·5535	0·1526	3·2005	3·3531	0·9545
39	4·0149	0·2491	1·8829	2·1320	0·8832	89	6·6194	0·1511	3·2342	3·3852	0·9554
1·40	4·0552	0·2466	1·9043	2·1509	0·8854	1·90	6·6859	0·1496	3·2682	3·4177	0·9562
41	4·0960	0·2441	1·9259	2·1701	0·8875	91	6·7531	0·1481	3·3025	3·4506	0·9571
42	4·1371	0·2417	1·9477	2·1894	0·8896	92	6·8210	0·1466	3·3372	3·4838	0·9579
43	4·1787	0·2393	1·9697	2·2090	0·8917	93	6·8895	0·1452	3·3722	3·5173	0·9587
44	4·2207	0·2369	1·9919	2·2288	0·8937	94	6·9588	0·1437	3·4075	3·5512	0·9595
45	4·2631	0·2346	2·0143	2·2488	0·8957	95	7·0287	0·1423	3·4432	3·5855	0·9603
46	4·3060	0·2322	2·0369	2·2691	0·8977	96	7·0993	0·1409	3·4792	3·6201	0·9611
47	4·3492	0·2299	2·0597	2·2896	0·8996	97	7·1707	0·1395	3·5156	3·6551	0·9619
48	4·3930	0·2276	2·0827	2·3103	0·9015	98	7·2427	0·1381	3·5523	3·6904	0·9626
49	4·4371	0·2254	2·1059	2·3312	0·9033	99	7·3155	0·1367	3·5894	3·7261	0·9633
1·50	4·4817	0·2231	2·1293	2·3524	0·9052	2·00	7·3891	0·1353	3·6269	3·7622	0·9640

HYPERBOLIC FUNCTIONS

x	e^x	e^{-x}	sinh$x$	cosh$x$	tanh$x$	x	e^x	e^{-x}	sinh$x$	cosh$x$	tanh$x$
2·01	7·4633	0·1340	3·6647	3·7987	0·9647	2·80	16·4447	0·0608	8·1919	8·2527	0·9926
02	7·5383	0·1327	3·7028	3·8355	0·9654	85	17·2878	0·0578	8·6150	8·6728	0·9933
03	7·6141	0·1313	3·7414	3·8727	0·9661	90	18·1742	0·0550	9·0596	9·1146	0·9940
04	7·6906	0·1300	3·7803	3·9103	0·9668	95	19·1060	0·0523	9·5268	9·5792	0·9945
05	7·7679	0·1287	3·8196	3·9483	0·9674	3·00	20·0855	0·0498	10·0179	10·0677	0·9951
06	7·8460	0·1275	3·8593	3·9867	0·9680	05	21·1153	0·0474	10·5340	10·5814	0·9955
07	7·9248	0·1262	3·8993	4·0255	0·9687	10	22·1980	0·0451	11·0765	11·1215	0·9960
08	8·0045	0·1249	3·9398	4·0647	0·9693	15	23·3361	0·0429	11·6466	11·6895	0·9963
09	8·0849	0·1237	3·9806	4·1043	0·9699	20	24·5325	0·0408	12·2459	12·2867	0·9967
2·10	8·1662	0·1225	4·0219	4·1443	0·9705	25	25·7903	0·0388	12·8758	12·9146	0·9970
11	3·2482	0·1212	4·0635	4·1847	0·9710	30	27·1126	0·0369	13·5379	13·5748	0·9973
12	8·3311	0·1200	4·1056	4·2256	0·9716	35	28·5027	0·0351	14·2338	14·2689	0·9975
13	3·4149	0·1188	4·1480	4·2669	0·9722	40	29·9641	0·0334	14·9654	14·9987	0·9978
14	8·4994	0·1177	4·1909	4·3086	0·9727	45	31·5004	0·0318	15·7343	15·7661	0·9980
15	8·5849	0·1167	4·2342	4·3507	0·9732	3·50	33·1155	0·0302	16·5426	16·5728	0·9982
16	8·6711	0·1153	4·2779	4·3932	0·9738	55	34·8133	0·0287	17·3923	17·4210	0·9984
17	8·7583	0·1142	4·3221	4·4362	0·9743	60	36·5982	0·0273	18·2855	18·3128	0·9985
18	8·8463	0·1130	4·3666	4·4797	0·9748	65	38·4747	0·0260	19·2243	19·2503	0·9987
19	8·9352	0·1119	4·4117	4·5236	0·9753	70	40·4473	0·0247	20·2113	20·2300	0·9988
2·20	9·0250	0·1108	4·4571	4·5679	0·9757	75	42·5211	0·0235	21·2488	21·2723	0·9989
21	9·1157	0·1097	4·5030	4·6127	0·9762	80	44·7012	0·0224	22·3394	22·3618	0·9990
22	9·2073	0·1086	4·5494	4·6580	0·9767	85	46·9931	0·0213	23·4859	23.5072	0·9991
23	9·2999	0·1075	4·5962	4·7037	0·9771	90	49·4025	0·0202	24·6911	24·7114	0·9992
24	9·3933	0·1065	4·6434	4·7499	0·9776	95	51·9354	0·0193	25·9581	25·9773	0·9993
25	9·4877	0·1054	4·6912	4·7966	0·9780	4·00	54·5982	0·0183	27·2899	27·3082	0·9993
26	9·5831	0·1044	4·7394	4·8437	0·9785	05	57·3975	0·0174	28·6900	28·7074	0·9994
27	9·6794	0·1033	4·7880	4·8914	0·9789	10	60·3403	0·0166	30·1619	30·1784	0·9995
28	9·7767	0·1023	4·8372	4·9395	0·9793	15	63·4340	0·0158	31·7091	31·7244	0·9995
29	9·8749	0·1013	4·8868	4·9881	0·9797	20	66·6863	0·0150	33·3357	33·3507	0·9996
2·30	9·9742	0·1003	4·9370	5·0372	0·9801	25	70·1054	0·0143	35·0456	35·0598	0·9996
31	10·0744	0·0993	4·9876	5·0868	0·9805	30	73·6998	0·0136	36·8431	36·8567	0·9996
32	10·1757	0·0983	5·0387	5·1370	0·9809	35	77·4785	0·0129	38·7328	38·7457	0·9997
33	10·2779	0·0973	5·0903	5·1876	0·9812	40	81·4509	0·0123	40·7193	40·7316	0·9997
34	10·3812	0·0963	5·1425	5·2388	0·9816	45	85·6269	0·0117	42·8076	42·8193	0·9997
35	10·4856	0·0954	5·1951	5·2905	0·9820	4·50	90·0171	0·0111	45·0030	45·0141	0·9998
36	10·5910	0·0944	5·2483	5·3427	0·9823	55	94·6324	0·0106	47·3109	47·3215	0·9998
37	10·6974	0·0935	5·3020	5·3954	0·9827	60	99·4843	0·0101	49·7371	49·7472	0·9998
38	10·8049	0·0926	5·3562	5·4487	0·9830	65	104·5850	0·0096	52·2877	52·2973	0·9998
39	10·9135	0·0916	5·4109	5·5026	0·9834	70	109·9472	0·0091	54·9690	54·9781	0·9998
2·40	11·0232	0·0907	5·4662	5·5570	0·9837	75	115·5843	0·0087	57·7878	57·7965	0·9999
41	11·1340	0·0898	5·5221	5·6119	0·9840	80	121·5104	0·0082	60·7511	60·7593	0·9999
42	11·2459	0·0889	5·5785	5·6674	0·9843	85	127·7404	0·0078	63·8663	63·8741	0·9999
43	11·3589	0·0880	5·6354	5·7235	0·9846	90	134·2898	0·0075	67·1412	67·1486	0·9999
44	11·4730	0·0872	5·6929	5·7801	0·9849	95	141·1750	0·0071	70·5839	70·5910	0·9999
45	11·5884	0·0863	5·7510	5·8373	0·9852	5·00	148·4132	0·0067	74·2032	74·2100	0·9999
46	11·7048	0·0854	5·8097	5·8951	0·9855	10	164·0219	0·0061	82·0079	82·0140	0·9999
47	11·8225	0·0846	5·8689	5·9535	0·9858	20	181·2722	0·0055	90·6334	90·6389	0·9999
48	11·9413	0·0837	5·9288	6·0125	0·9861	30	200·3368	0·0050	100·1659	100·1709	1·0000
49	12·0613	0·0829	5·9892	6·0721	0·9864	40	221·4064	0·0045	110·7010	110·7055	
2·50	12·1825	0·0821	6·0502	6·1323	0·9866	50	244·6919	0·0041	122·3439	122·3480	
55	12·8071	0·0781	6·3645	6·4426	0·9879	60	270·4264	0·0037	135·2114	135·2151	
60	13·4637	0·0743	6·6947	6·7690	0·9890	70	298·8674	0·0034	149·4320	149·4354	
65	14·1540	0·0707	7·0417	7·1123	0·9901	80	330·2996	0·0030	165·1483	165·1513	
70	14·8797	0·0672	7·4063	7·4735	0·9910	90	365·0375	0·0027	182·5174	182·5201	
75	15·6426	0·0639	7·7894	7·8533	0·9919	6·00	403·4288	0·0025	201·7132	201·7156	

DEGREES TO RADIAN

Degree	Radian	Degree	Radian	Degree	Radian	Degree	Radian	Degree	Radian	Degree	Radian
1	0·01745	31	0·54105	61	1·06465	91	1·58825	121	2·11185	151	2·63545
2	0·03491	32	0·55851	62	1·08210	92	1·60570	122	2·12930	152	2·65290
3	0·05236	33	0·57596	63	1·09956	93	1·62316	123	2·14675	153	2·67035
4	0·06981	34	0·59341	64	1·11701	94	1·64061	124	2·16421	154	2·68781
5	0·08727	35	0·61087	65	1·13446	95	1·65806	125	2·18166	155	2·70526
6	0·10472	36	0·62832	66	1·15192	96	1·67552	126	2·19911	156	2·72271
7	0·12217	37	0·64577	67	1·16937	97	1·69297	127	2·21657	157	2·74017
8	0·13963	38	0·66323	68	1·18682	98	1·71042	128	2·23402	158	2·75762
9	0·15708	39	0·68068	69	1·20428	99	1·72788	129	2·25147	159	2·77507
10	0·17453	40	0·69813	70	1·22173	100	1·74533	130	2·26893	160	2·79253
11	0·19199	41	0·71558	71	1·23918	101	1·76278	131	2·28638	161	2·80998
12	0·20944	42	0·73304	72	1·25664	102	1·78024	132	2·30383	162	2·82743
13	0·22689	43	0·75049	73	1·27409	103	1·79769	133	2·32129	163	2·84489
14	0·24435	44	0·76794	74	1·29154	104	1·81514	134	2·33874	164	2·86234
15	0·26180	45	0·78540	75	1·30900	105	1·83260	135	2·35619	165	2·87979
16	0·27925	46	0·80285	76	1·32645	106	1·85005	136	2·37365	166	2·89725
17	0·29671	47	0·82030	77	1·34390	107	1·86750	137	2·39110	167	2·91470
18	0·31416	48	0·83776	78	1·36136	108	1·88496	138	2·40855	168	2·93215
19	0·33161	49	0·85521	79	1·37881	109	1·90241	139	2·42601	169	2·94961
20	0·34907	50	0·87266	80	1·39626	110	1·91986	140	2·44346	170	2·96706
21	0·36652	51	0·89012	81	1·41372	111	1·93732	141	2·46091	171	2·98451
22	0·38397	52	0·90757	82	1·43117	112	1·95477	142	2·47837	172	3·00197
23	0·40143	53	0·92502	83	1·44862	113	1·97222	143	2·49582	173	3·01942
24	0·41888	54	0·94248	84	1·46608	114	1·98968	144	2·51327	174	3·03687
25	0·43633	55	0·95993	85	1·48353	115	2·00713	145	2·53073	175	3·05433
26	0·45379	56	0·97738	86	1·50098	116	2·02458	146	2·54818	176	3·07178
27	0·47124	57	0·99484	87	1·51844	117	2·04204	147	2·56563	177	3·08923
28	0·48869	58	1·01229	88	1·53589	118	2·05949	148	2·58309	178	3·10669
29	0·50615	59	1·02974	89	1·55334	119	2·07694	149	2·60054	179	3·12414
30	0·52360	60	1·04720	90	1·57080	120	2·09440	150	2·61799	180	3·14159

MINUTES TO RADIAN

Minutes	Radians	Minutes	Radians	Minutes	Radians	Minutes	Radians	Minutes	Radians
1	0·00029	13	0·00378	25	0·00727	37	0·01076	49	0·01425
2	058	14	407	26	0756	38	1105	50	1454
3	087	15	436	27	0785	39	1134	51	1484
4	116	16	465	28	0814	40	1164	52	1513
5	145	17	495	29	0844	41	1193	53	1542
6	175	18	524	30	0873	42	1222	54	1571
7	204	19	553	31	0902	43	1251	55	1600
8	233	20	582	32	0931	44	1280	56	1629
9	262	21	611	33	0960	45	1309	57	1658
10	291	22	640	34	0989	46	1338	58	1687
11	320	23	669	35	1018	47	1367	59	1716
12	0·00349	24	·00698	36	0·01047	48	0·01396	60	0·01745

SECONDS TO RADIAN

Seconds	Radians	Seconds	Radians	Seconds	Radians	Seconds	Radians	Seconds	Radians
1	0·00000	13	0·00006	25	0·00012	37	0·00018	49	0·00024
2	01	14	07	26	13	38	18	50	24
3	01	15	07	27	13	39	19	51	25
4	02	16	08	28	14	40	19	52	25
5	02	17	08	29	14	41	20	53	26
6	03	18	09	30	15	42	20	54	26
7	03	19	09	31	15	43	21	55	27
8	04	20	10	32	16	44	21	56	27
9	04	21	10	33	16	45	22	57	28
10	05	22	11	34	16	46	22	58	28
11	05	23	11	35	17	47	23	59	29
12	0·00006	24	0·00012	36	0·00017	48	0·00023	60	0·00029

Decimal equivalents of common fractions

Fractions	Decimals	Fractions	Decimals	Fractions	Decimals	Fractions	Decimals
1/64	0·0156	17/64	0·2656	33/64	0·5156	49/64	0·7656
1/32	0·0312	9/32	0·2813	17/32	0·5313	25/32	0·7813
3/64	0·0469	19/64	0·2969	35/64	0·5469	51/64	0·7969
1/16	0·0625	5/16	0·3125	9/16	0·5625	13/16	0·8125
5/64	0·0781	21/64	0·3281	37/64	0·5781	53/64	0·8281
3/32	0·0938	11/32	0·3438	19/32	0·5938	27/32	0·8438
7/64	0·1094	23/64	0·3594	39/64	0·6094	55/64	0·8594
1/8	0·1250	3/8	0·3750	5/8	0·6250	7/8	0·8750
9/64	0·1406	25/64	0·3906	41/64	0·6406	57/64	0·8906
5/32	0·1563	13/32	0·4063	21/32	0·6563	29/32	0·9063
11/64	0·1719	27/64	0·4219	43/64	0·6719	59/64	0·9219
3/16	0·1875	7/16	0·4375	11/16	0·6875	15/16	0·9375
13/64	0·2031	29/64	0·4531	45/64	0·7031	61/64	0·9531
7/32	0·2188	15/32	0·4688	23/32	0·7188	31/32	0·9688
15/64	0·2344	31/64	0·4844	47/64	0·7344	63/64	0·9844
1/4	0·2500	1/2	0·5000	3/4	0·7500	1	1·0000

Binomial coefficients

n	$\binom{n}{0}$	$\binom{n}{1}$	$\binom{n}{2}$	$\binom{n}{3}$	$\binom{n}{4}$	$\binom{n}{5}$	$\binom{n}{6}$	$\binom{n}{7}$	$\binom{n}{8}$	$\binom{n}{9}$	$\binom{n}{10}$
1	1	1									
2	1	2	1								
3	1	3	3	1							
4	1	4	6	4	1						
5	1	5	10	10	5	1					
6	1	6	15	20	15	6	1				
7	1	7	21	35	35	21	7	1			
8	1	8	28	56	70	56	28	8	1		
9	1	9	36	84	126	126	84	36	9	1	
10	1	10	45	120	210	252	210	120	45	10	1

Factorials

n	$n!$	$log\ n!$	n	$n!$	$log\ n!$
1	1	0·0000	6	720	2·8573
2	2	0·3010	7	5040	3·7024
3	6	0·7782	8	40320	4·6055
4	24	1·3802	9	362880	5·5598
5	120	2·0792	10	3628800	6·5598

BESSEL'S FUNCTIONS (Cylindrical)

x	$J_0(x)$	$J_1(x)$	$Y_0(x)$	$Y_1(x)$	$I_0(x)$	$I_1(x)$	$K_0(x)$	$K_1(x)$
0·0	1·0000	0·0000	−∞	−∞	1·0000	0·0000	∞	∞
0·1	0·9975	0·0499	−1·5342	−6·4590	1·003	0·0501	2·4271	9·8538
0·2	0·9900	0·0995	−1·0811	−3·3238	1·010	0·1005	1·7527	4·7760
0·3	0·9776	0·1483	−0·8073	−2·2931	1·023	0·1517	1·3725	3·0560
0·4	0·9604	0·1960	−0·6060	−1·7809	1·040	0·2040	1·1145	2·1844
0·5	0·9385	0·2423	−0·4445	−1·4715	1·063	0·2579	0·9244	1·6564
0·6	0·9120	0·2867	−0·3085	−1·2604	1·092	0·3137	0·7775	1·3028
0·7	0·8812	0·3290	−0·1907	−1·1032	1·126	0·3719	0·6605	1·0503
0·8	0·8463	0·3688	−0·0868	−0·9781	1·167	0·4329	0·5653	0·8618
0·9	0·8075	0·4059	0·0056	−0·8731	1·213	0·4971	0·4867	0·7165
1·0	0·7652	0·4401	0·0883	−0·7812	1·266	0·5652	0·4210	0·6019
1·1	0·7196	0·4709	0·1622	−0·6981	1·326	0·6375	0·3656	0·5098
1·2	0·6711	0·4983	0·2281	−0·6211	1·394	0·7147	0·3185	0·4346
1·3	0·6201	0·5220	0·2865	−0·5485	1·469	0·7973	0·2782	0·3725
1·4	0·5669	0·5419	0·3379	−0·4791	1·553	0·8861	0·2437	0·3208
1·5	0·5118	0·5579	0·3824	−0·4123	1·647	0·9817	0·2138	0·2774
1·6	0·4554	0·5699	0·4204	−0·3476	1·750	1·085	0·1880	0·2406
1·7	0·3980	0·5778	0·4520	−0·2847	1·864	1·196	0·1655	0·2094
1·8	0·3400	0·5815	0·4774	−0·2237	1·990	1·317	0·1459	0·1826
1·9	0·2818	0·5812	0·4968	−0·1644	2·128	1·448	0·1288	0·1597
2·0	0·2239	0·5767	0·5104	−0·1070	2·280	1·591	0·1139	0·1399
2·1	0·1666	0·5683	0·5183	−0·0577	2·446	1·745	0·1008	0·1227
2·2	0·1104	0·5560	0·5208	+0·0015	2·629	1·914	0·0893	0·1079
2·3	0·0555	0·5399	0·5181	0·0523	2·830	2·098	0·07914	0·09498
2·4	0·0025	0·5202	0·5104	0·1005	3·049	2·298	0·07022	0·08372
2·5	−0·0484	0·4971	0·4981	0·1459	3·290	2·517	0·06235	0·07389
2·6	−0·0968	0·4708	0·4813	0·1884	3·553	2·755	0·05540	0·06528
2·7	−0·1424	0·4416	0·4605	0·2276	3·842	3·016	0·04926	0·05774
2·8	−0·1850	0·4097	0·4359	0·2635	4·157	3·301	0·04382	0·05111
2·9	−0·2243	0·3754	0·4079	0·2959	4·503	3·613	0·03901	0·04529
3·0	−0·2601	0·3391	0·3769	0·3247	4·881	3·953	0·03474	0·04016
3·1	−0·2921	0·3009	0·3431	0·3496	5·294	6·326	0·03095	0·03563
3·2	−0·3202	0·2613	0·3070	0·3707	5·747	4·734	0·02759	0·03164
3·3	−0·3443	0·2207	0·2691	0·3879	6·243	5·181	0·02461	0·02812
3·4	−0·3643	0·1792	0·2296	0·4010	6·785	5·670	0·02196	0·02500
3·5	−0·3801	0·1374	0·1890	0·4102	7·378	6·206	0·01960	0·02224
3·6	−0·3918	0·0955	0·1477	0·4154	8·028	6·793	0·01750	0·01979
3·7	−0·3992	0·0538	0·1061	0·4167	8·739	7·436	0·01563	0·01763
3·8	−0·4026	0·0128	0·0645	0·4141	9·517	8·140	0·01397	0·01571
3·9	−0·4018	−0·0272	0·0234	0·4078	10·37	8·913	0·01248	0·01400
4·0	−0·3971	−0·0660	−0·0169	0·3979	11·30	9·759	0·01116	0·01248
4·1	−0·3887	−0·1033	−0·0561	0·3846	12·32	10·69	0·00998	0·01114
4·2	−0·3766	−0·1386	−0·0938	0·3680	13·44	11·71	0·00893	0·00994
4·3	−0·3610	−0·1719	−0·1296	0·3484	14·67	12·82	0·00799	0·00887
4·4	−0·3423	−0·2028	−0·1633	0·3260	16·01	14·05	0·00715	0·00792
4·5	−0·3205	−0·2311	−0·1947	0·3010	17·48	15·39	0·00640	0·00708
4·6	−0·2961	−0·2566	−0·2235	0·2737	19·09	16·86	0·00573	0·00633
4·7	−0·2693	−0·2791	−0·2494	0·2445	20·86	18·48	0·00513	0·00565
4·8	−0·2404	−0·2985	−0·2723	0·2136	22·79	20·25	0·00460	0·00506
4·9	−0·2097	−0·3147	−0·2921	0·1812	24·91	22·20	0·00412	0·00452
5·0	−0·1776	−0·3276	−0·3085	0·1479	27·24	24·34	0·00369	0·00404

BESSEL'S FUNCTIONS (Cylindrical)

x	$J_0(x)$	$J_1(x)$	$Y_0(x)$	$Y_1(x)$	$I_0(x)$	$I_1(x)$	$K_0(x)$ 0·00	$K_1(x)$ 0·00
5·1	—0·1443	—0·3371	—0·3216	0·1137	29·79	26·68	3308	3619
5·2	—0·1103	—0·3432	—0·3313	0·0792	32·58	29·25	2966	3239
5·3	—0·0758	—0·3460	—0·3374	0·0445	35·65	32·08	2659	2900
5·4	—0·0412	—0·3453	—0·3402	0·0101	39·01	35·18	2385	2597
5·5	—0·0068	—0·3414	—0·3395	—0·0238	42·69	38·59	2139	2326
5·6	0·0270	—0·3343	—0·3354	—0·0568	46·74	42·33	1918	2083
5·7	0·0599	—0·3241	—0·3282	—0·0887	51·17	46·44	1721	1806
5·8	0·0917	—0·3110	—0·3177	—0·1192	56·04	50·95	1544	1673
5·9	0·1220	—0·2951	—0·3044	—0·1481	61·38	55·90	1386	1499
6·0	0·1506	—0·2767	—0·2882	—0·1750	67·23	61·34	1244	1344
6·1	0·1773	—0·2559	—0·2694	—0·1998	73·66	67·32	1117	1205
6·2	0·2017	—0·2329	—0·2483	—0·2223	80·72	73·89	1003	1081
6·3	0·2238	—0·2081	—0·2251	—0·2422	88·46	81·10	09001	09691
6·4	0·2433	—0·1816	—0·1999	—0·2596	96·96	89·03	08083	08693
6·5	0·2601	—0·1538	—0·1732	—0·2741	106·3	97·74	07259	07799
6·6	0·2740	—0·1250	—0·1452	—0·2857	116·5	107·3	06520	06998
6·7	0·2851	—0·0953	—0·1162	—0·2945	127·8	117·8	05857	06280
6·8	0·2931	—0·0652	—0·0864	—0·3002	140·1	129·4	05262	05636
6·9	0·2981	—0·0349	—0·0563	—0·3029	153·7	142·1	04728	05659
7·0	0·3001	—0·0047	—0·0259	—0·3027	168·6	156·0	04248	04542
7·1	0·2991	0·0252	0·0042	—0·2995	185·0	171·4	03817	04078
7·2	0·2951	0·0543	0·0339	—0·2934	202·9	188·3	03431	03662
7·3	0·2882	0·0826	0·0628	—0·2846	222·7	206·8	03084	03288
7·4	0·2786	0·1096	0·0907	—0·2731	244·3	227·2	02772	02953
7·5	0·2663	0·1352	0·1173	—0·2591	268·2	249·6	02492	02653
7·6	0·2516	0·1592	0·1424	—0·2428	294·3	274·2	02240	02383
7·7	0·2346	0·1813	0·1658	—0·2243	323·1	301·3	02014	02141
7·8	0·2154	0·2014	0·1872	—0·2039	354·7	331·1	01811	01924
7·9	0·1944	0·2192	0·2065	—0·1817	389·4	363·9	01629	01729
8·0	0·1717	0·2346	0·2235	—0·1581	427·6	399·9	01465	01554
8·1	0·1475	0·2476	0·2381	—0·1331	469·3	439·5	01317	01396
8·2	0·1222	0·2580	0·2501	—0·1072	515·6	483·0	01185	01255
8·3	0·0960	0·2657	0·2595	—0·0806	566·3	531·0	01066	01128
8·4	0·0692	0·2708	0·2662	—0·0535	621·9	583·7	009588	01014
8·5	0·0419	0·2731	0·2702	—0·0262	683·2	641·6	008626	009120
8·6	0·0146	0·2728	0·2715	0·0011	750·5	705·4	007761	008200
8·7	—0·0125	0·2697	0·2700	0·0280	824·4	775·5	006983	007374
8·8	—0·0392	0·2641	0·2659	0·0544	905·8	852·7	006283	006631
8·9	—0·0653	0·2559	0·2592	0·0799	995·2	937·5	005654	005964
9·0	—0·0903	0·2453	0·2499	0·1043	1094	1031	005088	005364
9·1	—0·1142	0·2324	0·2383	0·1275	1202	1134	004579	004825
9·2	—0·1367	0·2174	0·2245	0·1491	1321	1247	004121	004340
9·3	—0·1577	0·2004	0·2086	0·1691	1451	1371	003710	003904
9·4	—0·1768	0·1816	0·1907	0·1871	1595	1508	003339	003512
9·5	—0·1939	0·1613	0·1712	0·2032	1753	1658	003006	003160
9·6	—0·2090	0·1395	0·1502	0·2171	1927	1824	002706	002843
9·7	—0·2218	0·1166	0·1279	0·2287	2119	2006	002436	002559
9·8	—0·2323	0·0928	0·1045	0·2379	2329	2207	002193	002302
9·9	—0·2403	0·0684	0·0804	0·2447	2561	2428	001975	002072
10·0	—0·2459	0·0435	0·0557	0·2490	2816	2671	001778	001865

AMOUNT 1 AT COMPOUND INTEREST:

$$s = (1 + r)^n$$

n	1%	2%	2½%	3%	3½%	4%	4½%	5%	6%
1	1.0100	1.0200	1.0250	1.0300	1.0350	1.0400	1.0450	1.0500	1.0600
2	1.0201	1.0404	1.0506	1.0609	1.0712	1.0816	1.0920	1.1025	1.1236
3	1.0303	1.0612	1.0769	1.0927	1.1087	1.1249	1.1412	1.1576	1.1910
4	1.0406	1.0824	1.1038	1.1255	1.1475	1.1699	1.1925	1.2155	1.2625
5	1.0510	1.1041	1.1314	1.1593	1.1877	1.2167	1.2462	1.2763	1.3382
6	1.0615	1.1262	1.1597	1.1941	1.2293	1.2653	1.3023	1.3401	1.4185
7	1.0721	1.1487	1.1887	1.2299	1.2723	1.3159	1.3609	1.4071	1.5036
8	1.0829	1.1717	1.2184	1.2668	1.3168	1.3686	1.4221	1.4775	1.5938
9	1.0937	1.1951	1.2489	1.3048	1.3629	1.4233	1.4861	1.5513	1.6895
10	1.1046	1.2190	1.2801	1.3439	1.4106	1.4802	1.5530	1.6289	1.7908
11	1.1157	1.2434	1.3121	1.3842	1.4600	1.5395	1.6229	1.7103	1.8983
12	1.1268	1.2682	1.3449	1.4258	1.5111	1.6010	1.6959	1.7959	2.0122
13	1.1381	1.2936	1.3785	1.4685	1.5640	1.6651	1.7722	1.8856	2.1329
14	1.1495	1.3195	1.4130	1.5126	1.6187	1.7317	1.8519	1.9799	2.2609
15	1.1610	1.3459	1.4483	1.5580	1.6753	1.8009	1.9353	2.0789	2.3966
16	1.1726	1.3728	1.4845	1.6047	1.7340	1.8730	2.0224	2.1829	2.5404
17	1.1843	1.4002	1.5216	1.6528	1.7947	1.9479	2.1134	2.2920	2.6928
18	1.1961	1.4282	1.5597	1.7024	1.8575	2.0258	2.2085	2.4066	2.8543
19	1.2081	1.4568	1.5987	1.7535	1.9225	2.1068	2.3079	2.5270	3.0256
20	1.2202	1.4859	1.6386	1.8061	1.9898	2.1911	2.4117	2.6533	3.2071
21	1.2324	1.5157	1.6796	1.8603	2.0594	2.2788	2.5202	2.7860	3.3996
22	1.2447	1.5460	1.7216	1.9161	2.1315	2.3699	2.6337	2.9253	3.6035
23	1.2572	1.5769	1.7646	1.9736	2.2061	2.4647	2.7522	3.0715	3.8197
24	1.2697	1.6084	1.8087	2.0328	2.2833	2.5633	2.8760	3.2251	4.0489
25	1.2824	1.6406	1.8539	2.0938	2.3632	2.6658	3.0054	3.3864	4.2919
26	1.2953	1.6734	1.9003	2.1566	2.4460	2.7725	3.1407	3.5557	4.5494
27	1.3082	1.7069	1.9478	2.2213	2.5316	2.8834	3.2820	3.7335	4.8223
28	1.3213	1.7410	1.9965	2.2879	2.6202	2.9987	3.4297	3.9201	5.1117
29	1.3345	1.7758	2.0464	2.3566	2.7119	3.1187	3.5840	4.1161	5.4184
30	1.3478	1.8114	2.0976	2.4273	2.8068	3.2434	3.7453	4.3219	5.7435
31	1.3613	1.8476	2.1500	2.5001	2.9050	3.3731	3.9139	4.5380	6.0881
32	1.3749	1.8845	2.2038	2.5751	3.0067	3.5081	4.0900	4.7649	6.4534
33	1.3887	1.9222	2.2589	2.6523	3.1119	3.6484	4.2740	5.0032	6.8406
34	1.4026	1.9607	2.3153	2.7319	3.2209	3.7943	4.4664	5.2533	7.2510
35	1.4166	1.9999	2.3732	2.8139	3.3336	3.9461	4.6673	5.5160	7.6861
36	1.4308	2.0399	2.4325	2.8983	3.4503	4.1039	4.8774	5.7918	8.1473
37	1.4451	2.0807	2.4933	2.9852	3.5710	4.2681	5.0969	6.0814	8.6361
38	1.4595	2.1223	2.5557	3.0748	3.6960	4.4388	5.3262	6.3855	9.1543
39	1.4741	2.1647	2.6196	3.1670	3.8254	4.6164	5.5659	6.7048	9.7035
40	1.4889	2.2080	2.6851	3.2620	3.9593	4.8010	5.8164	7.0400	10.2857
41	1.5038	2.2522	2.7522	3.3599	4.0978	4.9931	6.0781	7.3920	10.9029
42	1.5188	2.2972	2.8210	3.4607	4.2413	5.1928	6.3516	7.7616	11.5570
43	1.5340	2.3432	2.8915	3.5645	4.3897	5.4005	6.6374	8.1497	12.2505
44	1.5493	2.3901	2.9638	3.6715	4.5433	5.6165	6.9361	8.5572	12.9855
45	1.5648	2.4379	3.0379	3.7816	4.7024	5.8412	7.2482	8.9850	13.7646
46	1.5805	2.4866	3.1139	3.8950	4.8669	6.0748	7.5744	9.4343	14.5905
47	1.5963	2.5363	3.1917	4.0119	5.0373	6.3178	7.9153	9.9060	15.4659
48	1.6122	2.5871	3.2715	4.1323	5.2136	6.5705	8.2715	10.4013	16.3939
49	1.6283	2.6388	3.3533	4.2562	5.3961	6.8333	8.6437	10.9213	17.3775
50	1.6446	2.6916	3.4371	4.3839	5.5849	7.1067	9.0326	11.4674	18.4202

Note: A principal of 1 placed at a rate of interest r (expressed as decimal), compounded annually, accumulates to an amount $(1+r)^n$ at end of n years. The rate r is in per cent.

PRESENT VALUE OF 1 AT COMPOUND INTEREST:

$$v^n = (1 + r)^{-n}$$

n	1%	2%	2½%	3%	3½%	4%	4½%	5%	6%
1	.99010	.98039	.97561	.97087	.96618	.96154	.95694	.95238	.94340
2	.98030	.96117	.95181	.94260	.93351	.92456	.91573	.90703	.89000
3	.97059	.94232	.92860	.91514	.90194	.88900	.87630	.86384	.83962
4	.96098	.92385	.90595	.88849	.87144	.85480	.83856	.82270	.79209
5	.95147	.90573	.88385	.86261	.84197	.82193	.80245	.78353	.74726
6	.94205	.88797	.86230	.83748	.81350	.79031	.76790	.74622	.70496
7	.93272	.87056	.84127	.81309	.78599	.75992	.73483	.71068	.66506
8	.92348	.85349	.82075	.78941	.75941	.73069	.70319	.67684	.62741
9	.91434	.83676	.80073	.76642	.73373	.70259	.67290	.64461	.59190
10	.90529	.82035	.78120	.74409	.70892	.67556	.64393	.61391	.55839
11	.89632	.80426	.76214	.72242	.68495	.64958	.61620	.58468	.52679
12	.88745	.78849	.74356	.70138	.66178	.62460	.58966	.55684	.49697
13	.87866	.77303	.72542	.68095	.63940	.60057	.56427	.53032	.46884
14	.86996	.75788	.70773	.66112	.61778	.57748	.53997	.50507	.44230
15	.86135	.74301	.69047	.64186	.59689	.55526	.51672	.48102	.41727
16	.85282	.72845	.67362	.62317	.57671	.53391	.49447	.45811	.39365
17	.84438	.71416	.65720	.60502	.55720	.51337	.47318	.43630	.37136
18	.83602	.70016	.64117	.58739	.53836	.49363	.45280	.41552	.35034
19	.82774	.68643	.62553	.57029	.52016	.47464	.43330	.39573	.33051
20	.81954	.67297	.61027	.55368	.50257	.45639	.41464	.37689	.31180
21	.81143	.65978	.59539	.53755	.48557	.43883	.39679	.35894	.29416
22	.80340	.64684	.58086	.52189	.46915	.42196	.37970	.34185	.27751
23	.79544	.63416	.56670	.50669	.45329	.40573	.36335	.32557	.26180
24	.78757	.62172	.55288	.49193	.43796	.39012	.34770	.31007	.24698
25	.77977	.60953	.53939	.47761	.42315	.37512	.33273	.29530	.23300
26	.77205	.59758	.52623	.46369	.40884	.36069	.31840	.28124	.21981
27	.76440	.58586	.51340	.45019	.39501	.34682	.30469	.26785	.20737
28	.75684	.57437	.50088	.43708	.38165	.33348	.29157	.25509	.19563
29	.74934	.56311	.48866	.42435	.36875	.32065	.27902	.24295	.18456
30	.74192	.55207	.47674	.41199	.35628	.30832	.26700	.23138	.17411
31	.73458	.54125	.46511	.39999	.34423	.29646	.25550	.22036	.16425
32	.72730	.53063	.45377	.38834	.33259	.28506	.24450	.20987	.15496
33	.72010	.52023	.44270	.37703	.32134	.27409	.23397	.19987	.14619
34	.71297	.51003	.43191	.36604	.31048	.26355	.22390	.19035	.13791
35	.70591	.50003	.42137	.35538	.29998	.25342	.21425	.18129	.13011
36	.69892	.49022	.41109	.34503	.28983	.24367	.20503	.17266	.12274
37	.69200	.48061	.40107	.33498	.28003	.23430	.19620	.16444	.11580
38	.68515	.47119	.39128	.32523	.27056	.22529	.18775	.15661	.10924
39	.67837	.46195	.38174	.31575	.26141	.21662	.17967	.14915	.10306
40	.67165	.45289	.37243	.30656	.25257	.20829	.17193	.14205	.09722
41	.66500	.44401	.36335	.29763	.24403	.20028	.16453	.13528	.09172
42	.65842	.43530	.35448	.28896	.23578	.19257	.15744	.12884	.08653
43	.65190	.42677	.34584	.28054	.22781	.18517	.15066	.12270	.08163
44	.64545	.41840	.33740	.27237	.22010	.17805	.14417	.11686	.07701
45	.63905	.41020	.32917	.26444	.21266	.17120	.13796	.11130	.07265
46	.63273	.40215	.32115	.25674	.20547	.16461	.13202	.10600	.06854
47	.62646	.39427	.31331	.24926	.19852	.15828	.12634	.10095	.06466
48	.62026	.38654	.30567	.24200	.19181	.15219	.12090	.09614	.06100
49	.61412	.37896	.29822	.23495	.18532	.14634	.11569	.09156	.05755
50	.60804	.37153	.29094	.22811	.17905	.14071	.11071	.08720	.05429

Note: The present quantity which in n years will accumulate to 1 at the rate of interest r (expressed as decimal), compounded annually, is $(1+n)^{-n}$. The rate r is in per cent.

AMOUNT OF AN ANNUITY OF 1:

$$s_{\overline{n}|} \text{ at } r = \frac{(1+r)^n - 1}{r}$$

n	1%	2%	2½%	3%	3½%	4%	4½%	5%	6%
1	1.0000	1.0000	1.0000	1.0000	1.0000	1.0000	1.0000	1.0000	1.0000
2	2.0100	2.0200	2.0250	2.0300	2.0350	2.0400	2.0450	2.0500	2.0600
3	3.0301	3.0604	3.0756	3.0909	3.1062	3.1216	3.1370	3.1525	3.1836
4	4.0604	4.1216	4.1525	4.1836	4.2149	4.2465	4.2782	4.3101	4.3746
5	5.1010	5.2040	5.2563	5.3091	5.3625	5.4163	5.4707	5.5256	5.6371
6	6.1520	6.3081	6.3877	6.4684	6.5502	6.6330	6.7169	6.8019	6.9753
7	7.2135	7.4343	7.5474	7.6625	7.7794	7.8983	8.0192	8.1420	8.3938
8	8.2857	8.5830	8.7361	8.8923	9.0517	9.2142	9.3800	9.5491	9.8975
9	9.3685	9.7546	9.9545	10.1591	10.3685	10.5828	10.8021	11.0266	11.4913
10	10.4622	10.9497	11.2034	11.4639	11.7314	12.0061	12.2882	12.5779	13.1808
11	11.5668	12.1687	12.4835	12.8078	13.1420	13.4864	13.8412	14.2068	14.9716
12	12.6825	13.4121	13.7956	14.1920	14.6020	15.0258	15.4640	15.9171	16.8699
13	13.8093	14.6803	15.1404	15.6178	16.1130	16.6268	17.1599	17.7130	18.8821
14	14.9474	15.9739	16.5190	17.0863	17.6770	18.2919	18.9321	19.5986	21.0151
15	16.0969	17.2934	17.9319	18.5989	19.2957	20.0236	20.7841	21.5786	23.2760
16	17.2579	18.6393	19.3802	20.1569	20.9710	21.8245	22.7193	23.6575	25.6725
17	18.4304	20.0121	20.8647	21.7616	22.7050	23.6975	24.7417	25.8404	28.2129
18	19.6147	21.4123	22.3863	23.4144	24.4997	25.6454	26.8551	28.1324	30.9057
19	20.8109	22.8406	23.9460	25.1169	26.3572	27.6712	29.0636	30.5390	33.7600
20	22.0190	24.2974	25.5447	26.8704	28.2797	29.7781	31.3714	33.0660	36.7856
21	23.2392	25.7833	27.1833	28.6765	30.2695	31.9692	33.7831	35.7193	39.9927
22	24.4716	27.2990	28.8629	30.5368	32.3289	34.2480	36.3034	38.5052	43.3923
23	25.7163	28.8450	30.5844	32.4529	34.4604	36.6179	38.9370	41.4305	46.9958
24	26.9735	30.4219	32.3490	34.4265	36.6665	39.0826	41.6892	44.5020	50.8156
25	28.2432	32.0303	34.1578	36.4593	38.9499	41.6459	44.5652	47.7271	54.8645
26	29.5256	33.6709	36.0117	38.5530	41.3131	44.3117	47.5706	51.1135	59.1564
27	30.8209	35.3443	37.9120	40.7096	43.7591	47.0842	50.7113	54.6691	63.7058
28	32.1291	37.0512	39.8598	42.9309	46.2906	49.9676	53.9933	58.4026	68.5231
29	33.4504	38.7922	41.8563	45.2189	48.9108	52.9663	57.4230	62.3227	73.6398
30	34.7849	40.5681	43.9027	47.5754	51.6227	56.0849	61.0071	66.4388	79.0582
31	36.1327	42.3794	46.0003	50.0027	54.4295	59.3283	64.7524	70.7608	84.8017
32	37.4941	44.2270	48.1503	52.5028	57.3345	62.7015	68.6662	75.2988	90.8898
33	38.8690	46.1116	50.3540	55.0778	60.3412	66.2095	72.7562	80.0638	97.3432
34	40.2577	48.0338	52.6129	57.7302	63.4532	69.8579	77.0303	85.0670	104.1838
35	41.6603	49.9945	54.9282	60.4621	66.6740	73.6522	81.4966	90.3203	111.4348
36	43.0769	51.9944	57.3014	63.2759	70.0076	77.5983	86.1640	95.8363	119.1209
37	44.5076	54.0343	59.7339	66.1742	73.4579	81.7022	91.0413	101.6281	127.2681
38	45.9527	56.1149	62.2273	69.1594	77.0289	85.9703	96.1382	107.7095	135.9042
39	47.4123	58.2372	64.7830	72.2342	80.7249	90.4091	101.4644	114.0950	145.0585
40	48.8864	60.4020	67.4026	75.4013	84.5503	95.0255	107.0303	120.7998	154.7620
41	50.3752	62.6100	70.0876	78.6633	88.5095	99.8265	112.8467	127.8398	165.0477
42	51.8790	64.8622	72.8398	82.0232	92.6074	104.8196	118.9248	135.2318	175.9505
43	53.3978	67.1595	75.[illegible]608	85.4839	96.8486	110.0124	125.2764	142.9933	187.5076
44	54.9318	69.5027	78.5523	89.0484	101.2383	115.4129	131.9138	151.1430	199.7580
45	56.4811	71.8927	81.5161	92.7199	105.7817	121.0294	138.8500	159.7002	212.7435
46	58.0459	74.3306	84.5540	96.5015	110.4840	126.8706	146.0982	168.6852	226.5081
47	59.6263	76.8172	87.6679	100.3965	115.3510	132.9454	153.6726	178.1194	241.0986
48	61.2226	79.3535	90.8596	104.4084	120.3883	139.2632	161.5879	188.0254	256.5645
49	62.8348	81.9406	94.1311	108.5406	125.6018	145.8337	169.8594	198.4267	272.9584
50	64.4632	84.5794	97.4843	112.7969	130.9979	152.6671	178.5030	209.3480	290.3359

Note: If 1 is deposited at the end of each successive year and interest at rate r (expressed as decimal), compounded annually, is paid on the accumulated deposit at the end of each year, the total amount accumulated at the end of n years is $[(1+r)^n - 1]/r$. The rate r is expressed in per cent.

PRESENT VALUE OF AN ANNUITY OF 1

$$a_{\overline{n}|} \text{ at } r = \frac{1-(1+r)^{-n}}{r}$$

n	1%	2%	2½%	3%	3½%	4%	4½%	5%	6%
1	0.9901	.9804	.9756	.9709	.9662	.9615	.9569	.9524	.9434
2	1.9704	1.9416	1.9274	1.9135	1.8997	1.8861	1.8727	1.8594	1.8334
3	2.9410	2.8839	2.8560	2.8286	2.8016	2.7751	2.7490	2.7232	2.6730
4	3.9020	3.8077	3.7620	3.7171	3.6731	3.6299	3.5875	3.5460	3.4651
5	4.8534	4.7135	4.6458	4.5797	4.5151	4.4518	4.3900	4.3295	4.2124
6	5.7955	5.6014	5.5081	5.4172	5.3286	5.2421	5.1579	5.0757	4.9173
7	6.7282	6.4720	6.3494	6.2303	6.1145	6.0021	5.8927	5.7864	5.5824
8	7.6517	7.3255	7.1701	7.0197	6.8740	6.7327	6.5959	6.4632	6.2098
9	8.5660	8.1622	7.9709	7.7861	7.6077	7.4353	7.2688	7.1078	6.8017
10	9.4713	8.9826	8.7521	8.5302	8.3166	8.1109	7.9127	7.7217	7.3601
11	10.3676	9.7868	9.5142	9.2526	9.0016	8.7605	8.5289	8.3064	7.8869
12	11.2551	10.5753	10.2578	9.9540	9.6633	9.3851	9.1186	8.8633	8.3838
13	12.1337	11.3484	10.9832	10.6350	10.3027	9.9856	9.6829	9.3936	8.8527
14	13.0037	12.1062	11.6909	11.2961	10.9205	10.5631	10.2228	9.8986	9.2950
15	13.8651	12.8493	12.3814	11.9379	11.5174	11.1184	10.7395	10.3797	9.7122
16	14.7179	13.5777	13.0550	12.5611	12.0941	11.6523	11.2340	10.8378	10.1059
17	15.5623	14.2919	13.7122	13.1661	12.6513	12.1657	11.7072	11.2741	10.4773
18	16.3983	14.9920	14.3534	13.7535	13.1897	12.6593	12.1600	11.6896	10.8276
19	17.2260	15.6785	14.9789	14.3238	13.7098	13.1339	12.5933	12.0853	11.1581
20	18.0456	16.3514	15.5892	14.8775	14.2124	13.5903	13.0079	12.4622	11.4699
21	18.8570	17.0112	16.1845	15.4150	14.6980	14.0292	13.4047	12.8212	11.7641
22	19.6604	17.6580	16.7654	15.9369	15.1671	14.4511	13.7844	13.1630	12.0416
23	20.4558	18.2922	17.3321	16.4436	15.6204	14.8568	14.1478	13.4886	12.3034
24	21.2434	18.9139	17.8850	16.9355	16.0584	15.2470	14.4955	13.7986	12.5504
25	22.0232	19.5235	18.4244	17.4131	16.4815	15.6221	14.8282	14.0939	12.7834
26	22.7952	20.1210	18.9506	17.8768	16.8904	15.9828	15.1466	14.3752	13.0032
27	23.5596	20.7069	19.4640	18.3270	17.2854	16.3296	15.4513	14.6430	13.2105
28	24.3164	21.2813	19.9649	18.7641	17.6670	16.6631	15.7429	14.8981	13.4062
29	25.0658	21.8444	20.4535	19.1885	18.0358	16.9837	16.0219	15.1411	13.5907
30	25.8077	22.3965	20.9303	19.6004	18.3920	17.2920	16.2889	15.3725	13.7648
31	26.5423	22.9377	21.3954	20.0004	18.7363	17.5885	16.5444	15.5928	13.9291
32	27.2696	23.4683	21.8492	20.3888	19.0689	17.8736	16.7889	15.8027	14.0840
33	27.9897	23.9886	22.2919	20.7658	19.3902	18.1476	17.0229	16.0025	14.2302
34	28.7027	24.4986	22.7238	21.1318	19.7007	18.4112	17.2468	16.1929	14.3681
35	29.4086	24.9986	23.1452	21.4872	20.0007	18.6646	17.4610	16.3742	14.4982
36	30.1075	25.4888	23.5563	21.8323	20.2905	18.9083	17.6660	16.5469	14.6210
37	30.7995	25.9695	23.9573	22.1672	20.5705	19.1426	17.8622	16.7113	14.7368
38	31.4847	26.4406	24.3486	22.4925	20.8411	19.3679	18.0500	16.8679	14.8460
39	32.1630	26.9026	24.7303	22.8082	21.1025	19.5845	18.2297	17.0170	14.9491
40	32.8347	27.3555	25.1028	23.1148	21.3551	19.7928	18.4016	17.1591	15.0463
41	33.4997	27.7995	25.4661	23.4124	21.5991	19.9931	18.5661	17.2944	15.1380
42	34.1581	28.2348	25.8206	23.7014	21.8349	20.1856	18.7235	17.4232	15.2245
43	34.8100	28.6616	26.1664	23.9819	22.0627	20.3708	18.8742	17.5459	15.3062
44	35.4555	29.0800	26.5038	24.2543	22.2828	20.5488	19.0184	17.6628	15.3832
45	36.0945	29.4902	26.8330	24.5187	22.4955	20.7200	19.1563	17.7741	15.4558
46	36.7272	29.8923	27.1542	24.7754	22.7009	20.8847	19.2884	17.8801	15.5244
47	37.3537	30.2866	27.4675	25.0247	22.8994	21.0429	19.4147	17.9810	15.5890
48	37.9740	30.6731	27.7732	25.2667	23.0912	21.1951	19.5356	18.0772	15.6500
49	38.5881	31.0521	28.0714	25.5017	23.2766	21.3415	19.6513	18.1687	15.7076
50	39.1961	31.4236	28.3623	25.7298	23.4556	21.4822	19.7620	18.2559	15.7619

Note: The total present amount which will supply an annuity of 1 at the end of each year for n years, beginning one year hence, (assuming that in successive years the amount not yet paid out earns interest at rate r, compounded annually), is $[1-(1+r)^{-n}]/r$. The rate r is in per cent.

THE ANNUITY THAT 1 WILL PURCHASE

$$a_{\overline{n}|}^{-1} \text{ at } r = r/\left[1-(1+r)^{-n}\right]$$

n	1%	2%	2½%	3%	3½%	4%	4½%	5%	6%
1	1.01000	1.02000	1.02500	1.03000	1.03500	1.04000	1.04500	1.05000	1.06000
2	0.50751	0.51505	0.51883	0.52261	0.52640	0.53020	0.53400	0.53780	0.54544
3	.34002	.34675	.35014	.35353	.35693	.36035	.36377	.36721	.37411
4	.25628	.26262	.26582	.26903	.27225	.27549	.27874	.28201	.28859
5	.20604	.21216	.21525	.21835	.22148	.22463	.22779	.23097	.23740
6	.17255	.17853	.18155	.18460	.18767	.19076	.19388	.19702	.20336
7	.14863	.15451	.15750	.16051	.16354	.16661	.16970	.17282	.17914
8	.13069	.13651	.13947	.14246	.14548	.14853	.15161	.15472	.16104
9	.11674	.12252	.12546	.12843	.13145	.13449	.13757	.14069	.14702
10	.10558	.11133	.11426	.11723	.12024	.12329	.12638	.12950	.13587
11	.09645	.10218	.10511	.10808	.11109	.11415	.11725	.12039	.12679
12	.08885	.09456	.09749	.10046	.10348	.10655	.10967	.11283	.11928
13	.08241	.08812	.09105	.09403	.09706	.10014	.10328	.10646	.11296
14	.07690	.08260	.08554	.08853	.09157	.09467	.09782	.10102	.10758
15	.07212	.07783	.08077	.08377	.08683	.08994	.09311	.09634	.10296
16	.06794	.07365	.07660	.07961	.08268	.08582	.08902	.09227	.09895
17	.06426	.06997	.07293	.07595	.07904	.08220	.08542	.08870	.09544
18	.06098	.06670	.06967	.07271	.07582	.07899	.08224	.08555	.09236
19	.05805	.06378	.06676	.06981	.07294	.07614	.07941	.08275	.08962
20	.05542	.06116	.06415	.06722	.07036	.07358	.07688	.08024	.08718
21	.05303	.05878	.06179	.06487	.06804	.07128	.07460	.07800	.08500
22	.05086	.05663	.05965	.06275	.06593	.06920	.07255	.07597	.08305
23	.04889	.05467	.05770	.06081	.06402	.06731	.07068	.07414	.08128
24	.04707	.05287	.05591	.05905	.06227	.06559	.06899	.07247	.07968
25	.04541	.05122	.05428	.05743	.06067	.06401	.06744	.07095	.07823
26	.04387	.04970	.05277	.05594	.05921	.06257	.06602	.06956	.07690
27	.04245	.04829	.05138	.05456	.05785	.06124	.06472	.06829	.07570
28	.04112	.04699	.05009	.05329	.05660	.06001	.06352	.06712	.07459
29	.03990	.04578	.04889	.05211	.05545	.05888	.06241	.06605	.07358
30	.03875	.04465	.04778	.05102	.05437	.05783	.06139	.06505	.07265
31	.03768	.04360	.04674	.05000	.05337	.05686	.06044	.06413	.07179
32	.03667	.04261	.04577	.04905	.05244	.05595	.05956	.06328	.07100
33	.03573	.04169	.04486	.04816	.05157	.05510	.05874	.06249	.07027
34	.03484	.04082	.04401	.04732	.05076	.05431	.05798	.06176	.06960
35	.03400	.04000	.04321	.04654	.05000	.05358	.05727	.06107	.06897
36	.03321	.03923	.04245	.04580	.04928	.05289	.05661	.06043	.06839
37	.03247	.03851	.04174	.04511	.04861	.05224	.05598	.05984	.06786
38	.03176	.03782	.04107	.04446	.04798	.05163	.05540	.05928	.06736
39	.03109	.03717	.04044	.04384	.04739	.05106	.05486	.05876	.06689
40	.03046	.03656	.03984	.04326	.04683	.05052	.05434	.05828	.06646
41	.02985	.03597	.03927	.04271	.04630	.05002	.05386	.05782	.06606
42	.02928	.03542	.03873	.04219	.04580	.04954	.05341	.05739	.06568
43	.02873	.03489	.03822	.04170	.04533	.04909	.05298	.05699	.06533
44	.02820	.03439	.03773	.04123	.04488	.04866	.05258	.05662	.06501
45	.02771	.03391	.03727	.04079	.04445	.04826	.05220	.05626	.06470
46	.02723	.03345	.03683	.04036	.04405	.04788	.05184	.05593	.06441
47	.02677	.03302	.03641	.03996	.04367	.04752	.05151	.05561	.06415
48	.02633	.03260	.03601	.03958	.04331	.04718	.05119	.05532	.06390
49	.02591	.03220	.03562	.03921	.04296	.04686	.05089	.05504	.06366
50	.02551	.03182	.03526	.03887	.04263	.04655	.05060	.05478	.06344

One (1) now will purchase an annuity $r\,[1-(1+r)^{-n}]$ at the end of each year for n years, beginning one year hence (assuming that in successive years the amount not yet paid out earns interest at rate r, expressed as decimal, compounded annually). The rate r is in per cent.

The total amount accumulated at the end of n years is 1 if an annuity $s_{\overline{n}|}^{-1}$ is deposited at the end of each successive year (beginning one year hence), and interest a rate r, compounded annually, is paid on the accumulated deposit at the end of each year.

$$s_{\overline{n}|}^{-1} = a_{\overline{n}|}^{-1} - r.$$

PHYSICAL CONSTANTS

The Elements

Atomic No.	Symbol	Atomic weight Based on $C^{12}=12$	Atomic Diameter. A°	Ionisation Potential e.v.
1	H	1.00797	3.0	13.6
2	He	4.0026	...	24.6
3	Li	6.939	3.13	5.4
4	Be	9.0122	2.25	9.3
5	B	10.811	1.76	8.3
6	C	12.01115	1.54	11.3
7	N	14.0067	1.06	14.5
8	O	15.9994	1.48	13.6
9	F	18.9984	1.86	17.4
10	Ne	20.183	3.20	21.6
11	Na	22.9898	3.83	5.1
12	Mg	24.312	3.20	7.6
13	Al	26.9815	2.86	6.0
14	Si	28.086	2.34	8.1
15	P	30.9738	2.16	10.4
16	S	32.064	2.12	10.4
17	Cl	35.453	1.94	13.0
18	A	39.948	3.82	15.8
19	K	39.102	4.76	4.3
20	Ca	40.08	3.93	6.1
21	Sc	44.956	3.20	6.6
22	Ti	47.90	2.93	6.8
23	V	50.942	2.71	6.7
24	Cr	51.996	2.57	6.8
25	Mn	54.938	2.50	7.4
26	Fe	55.847	2.52	7.9
27	Co	58.9332	2.50	7.9
28	Ni	58.71	2.49	7.6
29	Cu	63.54	2.55	7.7
30	Zn	65.37	2.75	9.4
31	Ga	69.72	2.70	6.0
32	Ge	72.59	2.79	7.9
33	As	74.9216	2.50	9.8
34	Se	78.96	2.32	9.8
35	Br	79.909	2.26	11.8
36	Kr	83.80	4.00	14.0
37	Rb	85.47	5.02	4.2
38	Sr	87.62	4.29	5.7
39	Y	88.905	3.62	6.4
40	Zr	91.22	3.19	6.8

Atomic No.	Symbol	Atomic weight Based on $C^{12}=12$	Atomic Diameter. A°	Ionisation Potential e.v.
41	Nb	92.906	2.94	6.9
42	Mo	95.94	2.80	7.1
43	Te	(99)	...	7.2
44	Ru	101.07	2.67	7.4
45	Rh	102.905	2.70	7.5
46	Pd	106.4	2.75	8.3
47	Ag	107.870	2.88	7.6
48	Cd	112.40	3.04	9.0
49	In	114.82	3.14	5.8
50	Sn	118.69	3.16	7.3
51	Sb	121.751	3.23	8.6
52	Te	127.60	2.90	9.0
53	I	126.9044	2.70	10.4
54	Xe	131.30	4.40	12.1
55	Cs	132.905	5.40	3.9
56	Ba	137.34	4.48	5.2
57	La	138.91	3.74	5.6
58	Ce	140.12	3.64	...
59	Pr	140.907	3.65	...
60	Nd	144.24	3.63	6.3
61	Pm	(145)	...	...
62	Sm	150.35	...	5.6
63	Eu	151.96	4.08	5.7
64	Gd	157.25	3.59	6.2
65	Tb	158.924	3.54	...
66	Dy	162.50	3.54	...
67	Ho	164.93	3.52	...
68	Er	167.26	3.50	...
69	Tm	168.934	3.48	...
70	Yb	173.04	3.87	6.2

Atomic No.	Symbol.	Atomic weight Based on $C^{12}=12$	Atomic Diameter. A°	Ionisation Potential e.v.
71	Lu	174.97	3.47	6.2
72	Hf	178.49	3.17	5.5
73	Ta	180.948	2.94	7.7
74	W	183.85	2.82	8.0
75	Re	186.2	2.75	7.9
76	Os	190.2	2.70	8.7
77	Ir	192.2	2.71	9.2
78	Pt	195.09	2.76	9.0
79	Au	196.967	2.88	9.2
80	Hg	200.59	3.10	10.4
81	Tl	204.37	3.42	6.1
82	Pb	207.19	3.49	7.4
83	Bi	208.98	3.64	7.3
84	Po	(209)	...	8.4
85	At	(210)	...	...
86	Nt	(222)	...	10.7
87	Fr	(223)	...	...
88	Ra	226.05	...	...
89	Ac	(227)	...	...
90	Th	232.038	3.6	...
91	Pa	(231)	...	...
92	U	238.03	3.0	...
93	Np	(237)	...	...
94	Pu	(244)	...	...
95	Am	(243)	...	...
96	Cm	(247)	...	...
97	Bk	(247)	...	...
98	Cf	(251)	...	...
99	Es	(254)	...	...
100	Fm	(253)	...	...
101	Md	(256)	...	...
102	No	(253)	...	...
103	Lw	(257)	...	...

Properties of Elements

Name of Element									
1	2	3	4	5	6	7	8	9	10
Aluminium	Al	13	S	2.70	660	2060	0.215	0.50	0.382
Antimony	Sb	51	S	6.62	630.5	1380	0.049	0.04	.026
Argon	A	18	G	1.40×10^{-3}	−190	−186	0.125	0.4×10^{-4}	...
Arsenic	As	33	S	5.72	613	817	0.082	...	0.029
Barium	Ba	56	S	3.50	714	1640	0.068	...	0.016
Beryllium	Bc	4	S	1.86	1277	2770	0.450	0.380	0.250
Bismuth	Bi	83	S	9.80	271	1560	0.034	0.020	0.009
Boron	B	5	S	2.34	...	...	0.309	...	10-12
Bromine	Br	35	L	3.12	−7.2	58	0.110	...	10-18
Cadmium	Cd	48	S	8.65	321	765	0.055	0.220	0.146
Calcium	Ca	20	S	1.55	838	1440	0.149	0.300	0.218
Califorium	Cf	98	S	...	...	...	...	...	...
Carbon	C	6	S	2.26	3726	4830	0.165	0.057	0.7×10^{-3}
Cerium	Ce	58	S	6.80	795	3468	0.042	0.026	0.013
Cesium	Cs	55	L	1.90	28.8	690	0.052	...	0.053
Chlorine	Cl	17	G	3.22×10^{-3}	−101	−34.7	0.116	0.2×10^{-4}	...
Chromium	Cr	24	S	7.19	1875	2465	0.100	0.160	0.078
Cobalt	Co	27	S	8.70	1495	.2900	0.110	0.160	0.160
Columbium	Cb	41	S	8.40	2415	3300	0.065	0.125	0.080
Copper	Cu	29	S	8.96	1083	2340	0.092	0.940	0.593
Fluorine	F	9	G	1.69×10^{-3}	−219.6	−188.2	0.116	0.2×10^{-4}	...
Gadolinium	Gd	64	S	7.89	1312	3000	0.071	0.021	0.007
Gallium	Ga	31	L	5.91	29.8	2237	0.079	...	0.580
Germanium	Ge	32	S	5.32	937	2830	0.073	0.140	0.022
Gold	Au	79	S	19.30	1063	2600	0.031	0.70	0.420
Helium	He	2	G	0.129×10^{-3}	269.7	−268.9	1.25	0.33×10^{-3}	...
Hydrogen	H	1	G	0.090×10^{-3}	259.7	−252.7	3.45	0.42×10^{-3}	...
Indium	In	49	S	7.31	156.2	2000	0.057	0.057	0.111
Iodine	I	53	S	4.94	113.7	183	0.052	0.001	10^{-15}
Iridium	Ir	77	S	22.50	2454	5300	0.031	0.140	0.189
Iron	Fe	26	S	7.86	1536	3000	0.110	0.140	0.100
Krypton	Kr	36	G	2.6×10^{-3}	−157	−152	...	0.2×10^{-4}	...
Lanthanum	La	57	S	6.17	920	3470	0.045	0.033	0.017
Lead	Pb	82	S	11.4	327.4	1625	0.031	0.083	0.046
Lithium	Li	3	S	0.53	108.5	1330	0.790	0.170	0.108
Lutecium	Lu	71	S	9.84	1625	3327	0.037	...	0.015
Magnesium	Mg	12	S	0.97	650	1107	0.250	0.380	0.224
Manganese	Mn	25	S	7.43	1245	2150	0.115	...	0.054
Mercury	Hg	80	L	13.6	−38.9	357	0.033	0.020	0.011
Molybdenum	Mo	42	S	10.2	2610	5560	0.061	0.350	0.190
Neodymium	Nd	60	S	7.0	1024	3027	0.045	0.013	0.013
Neon	Ne	10	G	1.2×10^{-3}	−248.6	−246	...	0.11×10^{-3}	...
Neptunium	Np	93	S	19.5	637	...	...		...

Nickel	Ni	28	S	8.9	1453	2900	0.105	0.140	0.145
Niobium	Nb	41	S	8.4	2415	3300	0.065	0.125	0.080
Nitrogen	N	7	G	0.81×10^{-3}	−210	−195.8	0.247	0.6×10^{-4}	...
Osmium	Os	76	S	22.6	2700	5500	0.031	...	0.110
Oxygen	O	8	G	1.14×10^{-3}	−218.8	−183	0.218	0.6×10^{-4}	...
Palladium	Pd	46	S	12.0	1552	3980	0.058	0.170	0.093
Phosphorus	P	15	S	1.82	44.2	280	1.77	...	1.0
Platinum	Pt	78	S	21.40	1773	4330	0.032	0.170	0.095
Plutonium	Pu	94	S	19.50	637	...	...	...	...
Potassium	K	19	S	0.86	63.7	760	0.177	0.230	0.143
Praseodymium	Pr	59	S	6.77	935	3127	0.048	0.028	0.015
Radium	Ra	88	S	5.00	700				
Rhenium	Re	75	S	21.00	3180	5900	0.033	0.170	0.051
Rhodium	Rh	45	S	12.40	1966	4500	0.059	0.210	0.220
Rubidium	Rb	37	S	1.53	38.9	688	0.080	...	0.080
Ruthenium	Ru	44	S	12.20	2500	4900	0.057	...	0.100
Samarium	Sm	62	S	7.54	1072	1900	0.042	...	0.011
Scandium	Sc	21	S	3.00	1539	2730	0.130	0.015	0.015
Selenium	Se	34	S	4.79	217	685	0.084	0.5×10^{-4}	0.080
Silicon	Si	14	S	2.33	1410	2680	0.162	0.200	0.100
Silver	Ag	47	S	10.50	960.8	1950	0.056	0.980	0.616
Sodium	Na	11	S	0.97	97.8	892	0.295	0.320	0.218
Strontium	Sr	38	S	2.60	768	1380	0.176	...	0.043
Sulphur	S	16	S	2.07	113.0	4446	0.175	0.7×10^{-3}	10^{-23}
Tantalum	Ta	73	S	16.60	2996	5425	0.036	0.130	0.081
Tellurium	Te	52	S	6.24	450	990	0.047	0.014	10^{-6}
Terbium	Tb	65	S	8.27	1356	2800	0.044	...	0.009
Thallium	Tl	81	S	11.90	303	1457	0.031	0.093	0.055
Thorium	Th	90	S	11.70	1750	3850	0.034	...	0.055
Thulium	Tm	69	S	9.05	1545	1727	0.038	...	0.011
Tin	Sn	50	S	7.30	281.9	2270	0.054	0.160	0.088
Titanium	Ti	22	S	4.51	1668	3260	0.126	...	0.024
Tungsten	W	74	S	19.30	3410	5930	0.032	0.500	0.181
Uranium	U	92	S	19.07	1132	3818	0.028	0.064	0.034
Vanadium	V	23	S	6.10	1900	3450	0.126	...	0.024
Kenon	Xe	54	G	3.06×10^{-3}	108	111.9	...	0.0001	...
Ytterbium	Yb	70	S	0.98	824	1427	0.035	...	0.035
Yttrium	Y	39	S	4.47	1509	2927	0.071	0.035	0.019
Zinc	Zn	30	S	7.14	419.5	906	0.092	0.270	0.167
Ziroonium	Zr	40	S	6.49	1852	3580	0.066		0.024

Elestic Constants

Substance	Young's Modulus dyness/cm^2 $\times 10^{11}$	Rigidity Modulus dyness/cm^2 $\times 10^{11}$	Bulk Modulus dyness/cm^2 $\times 10^{11}$	Poisson' Ratio μ
Metals				
Aluminium	6.9-7.2	2.4-2.7	7.3-7.7	.33-.35
Copper	11.0-12.9	3.4-4.6	13.0-14.3	.25-.35
Gold	7.8-8.2	2.7-2.9	16.5-16.7	.14-.43
Iron (wrought)	19.0-22.0	7.7-8.3	14.6-16.0	.27-.29
Iron (cast)	10.0-13.0	3.5-5.3	9.5-9.7	.23-.31
Lead	1.5-1.7	0.5-0.7	4.0-5.0	.43-.46
Nickel	20.0-21.0	7.0-8.0	17.0-18.0	.29-.31
Silver	7.7-7.9	2,8-3.0	10.8-11.0	.37-.39
Steel (cast)	19.0-21.0	7.4-7.6	16.5-17.5	.27-.29
Steel (mild)	21.0-23.0	8.0-8.9	16.0-19.0	.25-.31
Tin	5.3-5.5	1.9-2.1	5.1-5.5	.32-.34
Tungsten	38.5-39.5	14.5-15.5	29.0-31.0	
Zinc	8.0-11.0	3.6-3.8	3.0-6.0	.20-.30
Alloys				
Brass	9.0-10.2	3.4-3.5	8.0-11.0	.30-.40
Bronze	8.0-11.0	3.6-3.8	9.0-9.6	.35-.37
Constantan	16.2-16.4	6.0-6.2	15.4-15.6	.31-.33
German Silver	11.0-11.6	4.3-4.7	14.9-15.1	.36-.38
Manganin	12.3-12.5	4.6-4.8	12.0-12.2	.32-.34
Phosphorbronze	11.9-12.1	4.3-4.5	—	.37-.39
Miscellaneous				
Glass (crown)	6.0-7.8	2.6-3.2	4.0-5.9	.20-.27
Glass (flint)	5.0-6.0	2.0-2.5	3.6-3.8	.22-.26
India rubber	.048-.052	.00016	—	.46-.49
Quartz fibre	5.2-5.4	2.9-3.1	1.4-1.6	—

Thermal Constants of Solids

Material	Coefficient of Linear expansion $\times 10^{-6}$	Specific heat, Cal/Grm	Melting point, °C	Boiling point, °C	Thermal conductivity, a	Latent heart of fusion, Cal/Grm
Aluminium	25.5	0.22	(659.7)	2057	0.49	77
Antimony	11.4	0.05	(630.5)	1380	0.042	39
Bees-wax	...	...	62	...	0.00006	42
Bismuth	13.3	0.03	271	1560	0.019	14
Brass	18.9	0.09	900*	...	0.26	...
Bronze	18.0	0.09	900*	...	...	...
Cadminum	29.8	0.05	(320.9)	767	0.222	13
Cardboard	...	...	...	...	0.0005	...
Chromium	8.2	0.10	1890	2480	...	32
Cobalt	12.3	0.11	(1495)	2900	...	58
Copper	16.7	0.09	(1083)	2336	0.91	43
Cork	...	0.49	...	...	0.00012	...
Ebonite	84.0	0.40	...	...	0.0004	...
Glass	9.3	0.16	1100	...	0.0025	...
Gold	13.9	0.03	1063	2600	0.70	16
Ice	51.0	0.50	0	100	0.0050	80
India rubber	70.4	0.38*	...	...	0.00045	...
Invar	0.9	0.12	1500	...	...	...
Iron (cast)	10.2	0.12	1100	...	0.114	28
Iron (wrought)	11.9	0.11	1535	3000	0.14	...
Lead	29.2	0.03	(327.4)	1620	0.08	6
Magensium	25.6	0.25	651	1107	0.376	37
Naphthalene	...	...	80	218	...	35
Nickel	12.8	0.11	(1455)	2900	0.14	73
Paraffin wax	...	0.69	55*	390*	0.0006	42
Phosphorus	125.0	0.20	44	280	...	5
Platinum	8.9	0.03	(1773.5)	43.00	0.166	27
Silica (fused)	.42	0.17	1700	...	0.0024	00
Silver	19.2	0.06	961	1950		22
Steel	11.0	0.11	14.00	...	0.11	...
Sulphur (Romb.)	70.0	0.17	113	444.6	0.00063	9
Tin	22.3	0.05	(231.9)	2270	0.15	15
Tungsten	4.0	0.03	(3370)	5900	0.48	...
Woods metal	...	0.04	65	...	...	8
Zinc	26.3	0.09	(419)	907	0.26	27

a. Calories per sec. per sq. cm. unit temperature gradient

* Liable to vary with different specimens.

Thermal constants of Liquids

	Coef. of Cub. Exp.	Sp. heat, Cal/gm.	Freezing Point, °C	Boiling Point, °C	Thermal conductivity a	Latent heat of vaporisation b	Criti tempe tur c
	$\times 10^{-4}$				$\times 10^{-8}$		
Alcohol (ethyl)	11.0	0.62	—117.3	78.1	0.42	205	24
" (methyl)	12.2	0.60	— 97.0	64.7	0.51	267	42
Aniline	8.5	0.51	— 7.0	184.2	0.43	104	42
Benzene	12.4	0.41	5.4	80.2	0.41	95	28
Bromine	11.7	0.11	7.3	58.0	...	46	30
Carbon disulphide	12.1	0.24	—111.6	46.0	0.38	84	27
Carbon tetrachloride	12.3	0.20	— 23.0	76.7	0.25	46	28
Castor oil	6.5	...	...	265	0.43	...	...
Chloroform	12.6	0.23	— 63.5	61.2	0.33	58	26
Cocoanut oil	7.4	0.49	25.0	286	...	...	...
Ether (ethyl)	16.3	0.56	—120.0	34.31	0.33	88	194
Glycerine	5.3	0.58	17.0	290	0.68	...	...
Kerosine	...	0.50	...	240	0.38	72	...
Mercury	1.8	0.03	(— 38.9)	(356.6)	19.00	65	...
Olive oil	6.3	0.47	...	...	0.42	...	...
Paraffin oil	9.0	0.52	...	...	0.35	...	...
Toluene	10.9	0.40	— 95.1	110.5	0.38	84	321
Turpentine	9.4	0.42	...	156	0.30	70	...
Water		1.00	(0)	(100)	1.40	539	374

a. Calories per sec. per sq. cm. per unit temperature gradient.
b. Calories per gm. at normal boiling point.
c. Degrees Centigrade.

Valencies and Electro-chemical Equivalents of Elements

Atomic No.	Symbol	Valency	e.c.e. milligrams per coulomb	Atomic No.	Symbol.	Valency	e.c.e. milligrams per coulomb	Atomic No.	Symbol.	Valency	e.c.e. milligrams per coulomb
1	H	1	0.0104	29	Cu	1	0.6590	58	Ce	3	0.4845
3	Li	1	0.0725			2	0.3295			4	0.3634
4	Be	2	0.0471	30	Zn	2	0.3387	59	Pr	3	0.4857
5	B	3	0.0380	31	Ga	3	0.2414	60	Nd	3	0.4985
6	C	4	0.0313	32	Ge	4	0.1877	62	Sm	3	0.5195
7	N	3	0.0484	33	As	3	0.2590	63	Eu	3	0.5251
		5	0.0290			5	0.1554	64	Gd	3	0.5434
8	O	2	0.0829	34	Se	2	0.4104	65	Tb	3	0.5499
9	F	1	0.1968	35	Br	1	0.8282	66	Dy	3	0.5613
11	Na	1	0.2384	37	Rb	1	0.8855	67	Ho	3	0.5648
12	Mg	2	0.1260	38	Sr	2	0.4540	68	Er	3	0.5793
13	Al	3	0.0932	39	Y	3	0.3074	69	Tm	3	0.5821
14	Si	4	0.0733	40	Zr	4	0.2347	70	Yb	3	0.5976
15	P	3	0.1071	41	Nb	5	0.1937	71	Lu	3	0.6045
		5	0.0642	42	Mo	4	0.2487	73	Ta	5	0.3751
16	S	2	0.1662			6	0.1658	74	W	6	0.3178
		4	0.0831	44	Ru	6	0.1757	76	Os	6	0.3297
17	Cl	1	0.3675			8	0.1317	77	Ir	4	0.5003
19	K	1	0.4052	45	Rh	3	0.3554	78	Pt	2	1.0104
20	Ca	2	0.2076	46	Pd	2	0.5528			4	0.5052
21	Sc	3	0.1523			4	0.2764	79	Au	1	2.0436
22	Ti	4	0.1246	47	Ag	1	1.1180			3	0.6812
23	V	3	0.1762	48	Cd	2	0.5824	80	Hg	1	2.0788
		5	0.1057	49	In	3	0.3966			2	1.0394
24	Cr	3	0.1800	50	Sn	2	0.6166	81	Tl	1	2.1141
		6	0.0900			4	0.3083	82	Pb	2	1.0736
25	Mn	2	0.2846	51	Sb	3	0.4152			4	0.5368
		3	0.1897			5	0.2491	83	Bi	3	0.7185
26	Fe	2	0.2893	52	Te	2	0.6606			5	0.4311
		3	0.1929	53	I	1	1.3150	88	Ra	2	1.1731
27	Co	2	0.3061	55	Cs	1	1.3764	90	Th	4	0.6021
		3	0.2041	56	Ba	2	0.7118	92	U	4	0.6179
28	Ni	2	0.3040	57	La	3	0.4802			6	0.4119
		3	0.2027								

Electrical resistances

Substance	ρ	α	Substance	ρ	α
Aluminium	2.82	.0036	Nichrome	110.00	.00035
Brass (wire)	6.60	.0016	Nickel	7.24	.0054
Constantan	42.40	.00017	Phosphor-bronze	4.00	.0035
Copper	1.72	.004	Platinoid	34.40	.00025
German silver	26.60	.00038	Platinum	10.50	.0037
Iron	9.80	.005	Silver	1.62	.0036
Lead	20.63	.004	Steel (piano wire)	12.00	.0032
Magnesium	4.46	.0039	Tin	11.40	.0044
Manganin	43.00	.000021	Tungsten	5.50	.0052
Mercury	95.77	.00089	Zinc	5.92	.0035

ρ Specific resistance in 10^{-6} ohm. cm. at 20° C.
α Increase in resistance per ohm. per 1° C. rise in temperature.

Resistance wires (S.W.G.)

S.W.G	Ohms per metre				Amperes at*		
	German Silver	Manganin	Platinoid	Eureka	100° C.	200° C.	300° C.
18	0.193	0.361	0.349	0.405	4.3	7.0	9.8
20	0.345	0.645	0.621	0.722	3.0	4.7	5.9
22	0.570	1.07	1.03	1.20	2.2	3.2	4.1
24	0.920	1.73	1.67	1.93	1.5	2.3	2.8
26	1.38	2.58	2.50	2.89	1.0	1.7	2.1
28	2.02	3.82	3.69	4.27	0.8	1.4	1.6
30	2.90	5.45	5.25	6.08	0.6	1.0	1.3
32	3.83	7.18	6.81	8.01	0.5	0.8	1.0
34	5.27	9.90	9.55	11.0	0.4	0.6	0.8
36	7.74	14.5	13.9	16.2	0.3	0.5	0.6
38	12.4	23.2	22.4	26.0	0.2	0.3	0.4
40	19.4	36.3	35.0	40.5	0.15	0.2	0.3

Nichrome wires (S.W.G.)

S.W.G	Ohms per metre	Amperes at				S.W.G	Ohms per metre	Amperes at			
		100° C.	300° C.	500° C.	1000° C.			100° C.	300° C.	500° C.	1000° C.
18	0.80	4.2	10.5	15.3	25.3	**30**	12.0	0.6	1.5	2.3	3.8
20	1.42	2.8	6.9	10.1	16.9	**32**	15.8	0.5	1.2	1.8	3.0
22	2.34	2.0	4.9	7.2	12.2	**34**	21.7	0.4	0.9	1.4	2.3
24	3.80	1.4	3.5	5.2	8.7	**36**	31.8	0.3	0.7	1.1	1.8
26	5.68	1.0	2.5	3.7	6.3	**38**	52.2	0.21	0.5	0.8	1.3
28	8.39	0.7	1.8	2.7	4.5	**40**	79.7	0.17	0.4	0.6	1.0

Copper wires (S.W.G.)

S.W.G	Ohms per 1,000 metres.	Safe current amps.	S.W.G	Ohms per 1,000 metres.	Safe current amps.	S.W.G	Ohms per 1,000 metres.	Safe current amps.	S.W.G	Ohms per 1,000 metres.	Safe current amps.
12	3.09	15.0	**22**	42.68	1.7	**32**	286.9	0.3	**42**	2091	0.04
14	5.23	9.8	**24**	69.13	1.1	**34**	395.9	0.2	**44**	3268	0.025
16	8.17	6.8	**26**	103.3	0.7	**36**	579.3	0.15	**46**	5806	0.015
18	14.52	4.2	**28**	152.8	0.5	**38**	929.4	0.10	**48**	13070	0.006
20	25.82	2.6	**30**	217.6	0.4	**40**	1452.0	0.06	**50**	33460	0.003

Insulators

Substance	Specific Resistance, ohm-cm	Substance	Specific Resistance, ohm-cm
Amber	5×10^{16}	Rubber (hard)	1×10^{18}
Bakelite	7.5×10^{15}	Shellac	1×10^{16}
Ebonite	2×10^{16}	Sulphur	1×10^{17}
Fiber	2×10^{9}	Soil (dry)	$>10^{6}$
Glass (plate)	2×10^{13}	Soil (damp)	5×10^{3}
Mica	2×10^{17}	Water	1×10^{6}
Paraffin	5×10^{16}	Castor oil	3×10^{18}
Porcelain	1×10^{14}	Linseed oil	5×10^{10}
Quartz (fused)	5×10^{18}	Olive oil	1×10^{12}

ρ Approximate specific resistance at room temperature in ohm. cm.

Dielectric constants (Vacuum = 1)

Substance	ϵr	Substance	ϵr	Substance	ϵr
Amber	2.8	Mica	7.0	Sulphur	3.9
Asbestos	3.0	Paper (dry)	2.4	Soil (dry)	2.9
Bakelite	4.5	Paper (waxed)	2.8	Soil (damp)	8.0
Bitumen	3.5	Paraffin	2.1	Aniline	7.2
Ebonite	2.7	Porcelain	7.0	Castor oil	4.5
Fibre	5.0	Quartz (fused)	4.0	Glycerine	43.0
Glass (plate)	7.6	Rubber (hard)	3.1	Linseed oil	8.5
Granite	8.0	Sealing wax	3.0	Olive oil	3.1
Gypsum	6.3	Shellac	2.9	Transformer oil	2.2
Marble	8.0	Steatite	6.0	Water	81.0

SOUND ABSORPTION COEFFICIENTS

(Ratio of absorbed intensity to the incident intensity)

Materials	at frequencies					
	125	250	500	1000	2000	4000
Asbestos spray	.10	.20	.30	.40	.50	.60
Brick wall	.04	.03	.03	.04	.05	.07
Coir matting	.10	.10	.20	.20	.30	.45
Concrete wall	.02	.02	.02	.04	.05	.05
False ceiling board	.30	.20	.20	.15	.15	.10
Felt	.15	.20	.50	.60	.70	.90
Fibre boards	.10	.20	.30	.40	.50	.50
Floor tiles	.02	.02	.03	.03	.04	.05
Glass window Pane	.30	.20	.10	.10	.05	.05
Heavy carpet	.07	.10	.30	.50	.55	.60
Heavy curtains plain	.10	.20	.30	.35	.40	.50
Heavy curtains draped	.15	.30	.50	.70	.70	.70
Light curtains—plain	.04	.10	.15	.20	.30	.30
Light curtains—draped	.06	.20	.40	.50	.50	.50
Lime or Gypsum plaster	.03	.03	.03	.03	.04	.05
Linoleum	.01	.01	.02	.02	.03	.05
Marble or galzed tile	.01	.01	.01	.01	.02	.02
Painted or distempered surface	.01	.01	.01	.02	.02	.03
Pitched roof (underside)	.15	.10	.10	.10	.15	.20
Wood Panelling	.25	.25	.20	.15	.10	.10
Wood-wool slabs	.10	.20	.40	.50	.60	.60

BRITISH-METRIC DIRECT CONVERSIONS

Direct Conversion of LENGTH

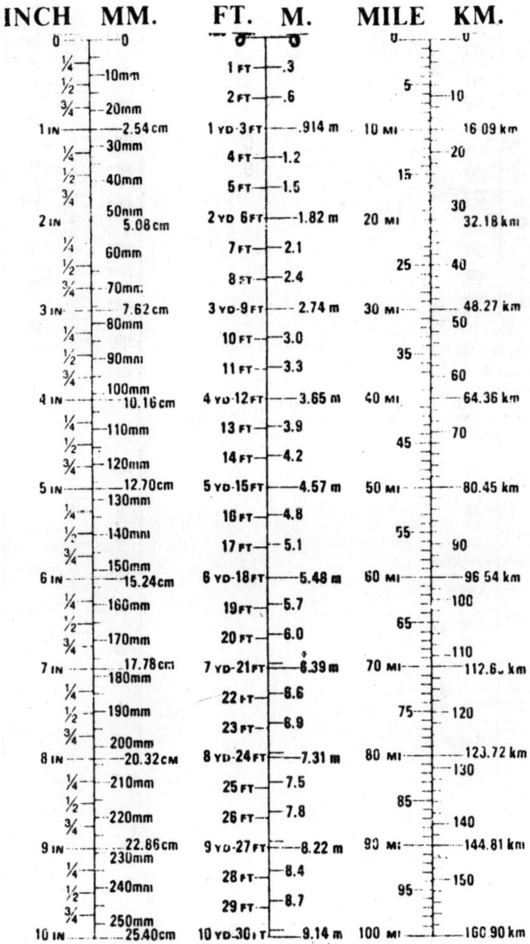

Direct Conversion

of AREA

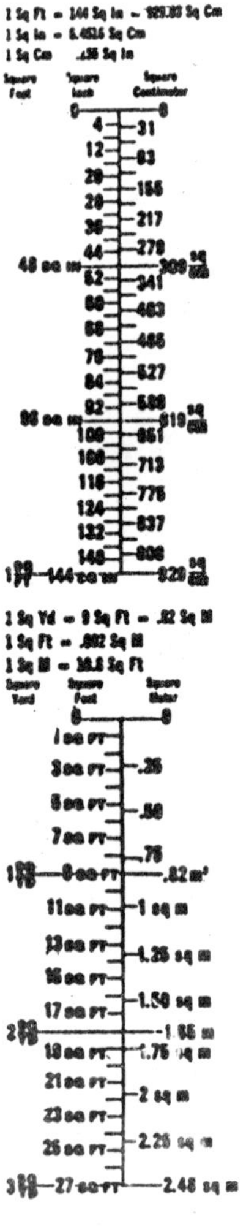

of TEMPERATURE

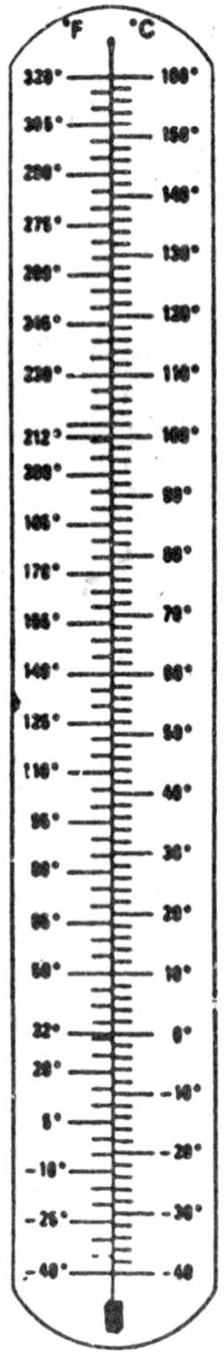

Direct Conversion of Flow and Volume

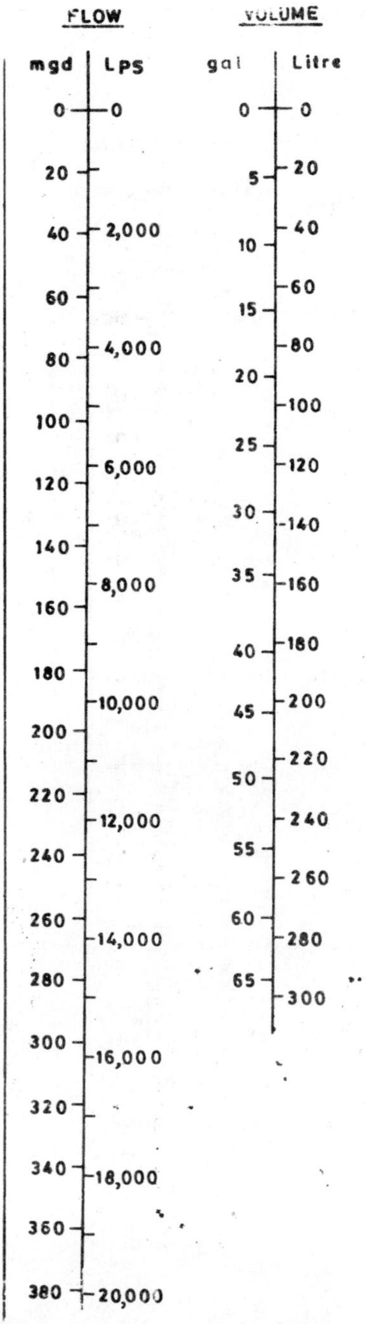

Direct Conversion of Population Density and Pipe Diameter

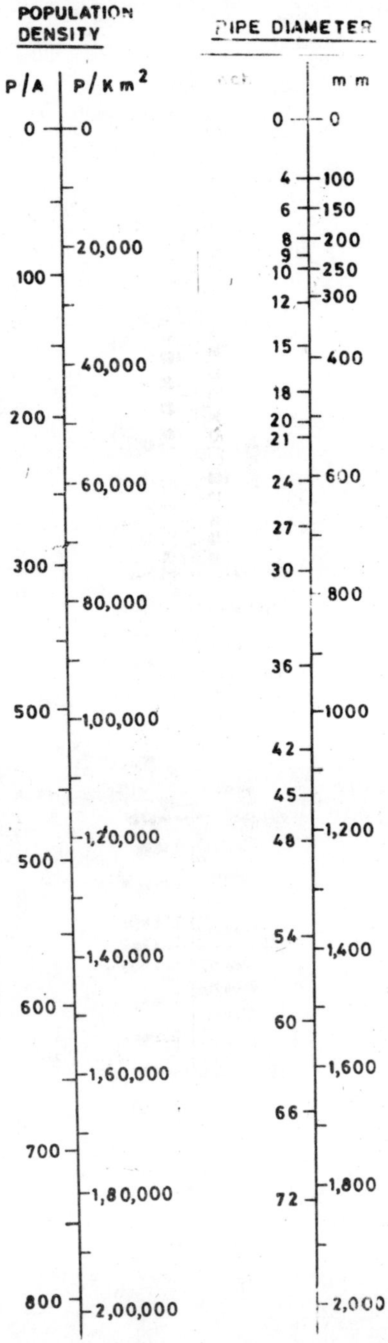

Direct Conversion of WEIGHT

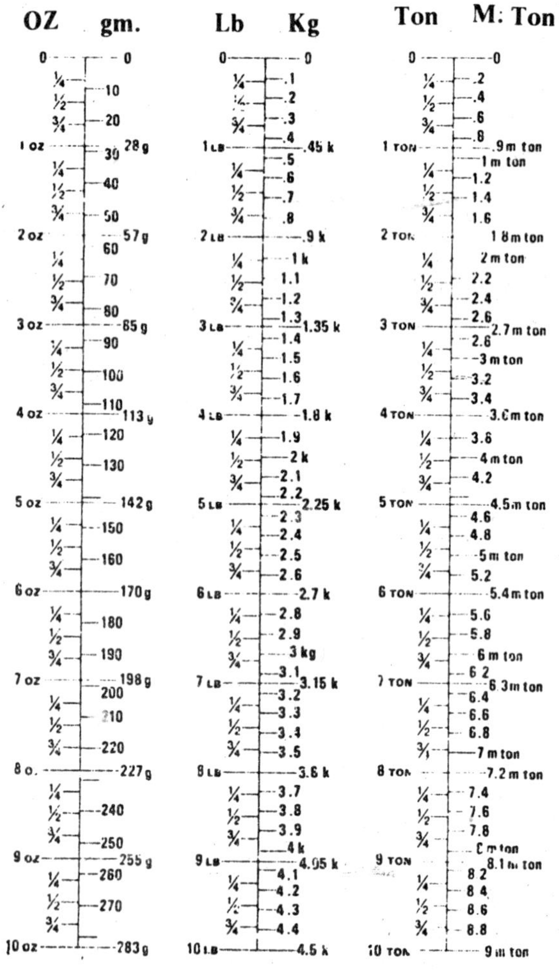